A Liquid Medium Energy Saving Window Technique and Application
液体介质节能窗技术与应用

吕原丽 著

Beijing
Metallurgical Industry Press
2020

Abstract

Liquid flow window is a novel multi-glazing system with liquid medium in the window cavity to serve as solar thermal collector and heat/cool radiator. Provided in the book is a comparative analysis of its thermal characteristics. Thermal performance optimizations considering the varied schematic design parameters, hot water demand patterns and climates are introduced. And economic, energy, and environmental life cycle assessment are also conducted to quantify the sustainability of the buoyant driven liquid flow window.

This book provides general knowledge about glazing techniques for researchers and engineers. And the systematic studies can also be references for researchers in the area of building integrated solar energy application.

图书在版编目(CIP)数据

液体介质节能窗技术与应用=A Liquid Medium Energy Saving Window Technique and Application/吕原丽著. —北京：冶金工业出版社，2020.8

ISBN 978-7-5024-6888-0

Ⅰ.①液… Ⅱ.①吕… Ⅲ.①窗—建筑设计—节能设计 Ⅳ.①TU228

中国版本图书馆 CIP 数据核字（2020）第 162032 号

出版人　陈玉千
地　　址　北京市东城区嵩祝院北巷 39 号　邮编　100009　电话　(010)64027926
网　　址　www.cnmip.com.cn　电子信箱　yjcbs@cnmip.com.cn
责任编辑　杜婷婷　刘林烨　美术编辑　郑小利　版式设计　孙跃红
责任校对　郑　娟　责任印制　李玉山
ISBN 978-7-5024-6888-0
冶金工业出版社出版发行；各地新华书店经销；三河市双峰印刷装订有限公司印刷
2020 年 8 月第 1 版，2020 年 8 月第 1 次印刷
169mm×239mm；9.75 印张；188 千字；146 页
59.00 元

冶金工业出版社　投稿电话　(010)64027932　投稿信箱　tougao@cnmip.com.cn
冶金工业出版社营销中心　电话　(010)64044283　传真　(010)64027893
冶金工业出版社天猫旗舰店　yjgycbs.tmall.com

(本书如有印装质量问题，本社营销中心负责退换)

Preface

The importance of building energy saving and sustainable building development has been put forward under the worldwide energy and environmental crisis. In mordern buildings, the glazed façade is widely used due to its features of esthetic appearance and wide vision. However, it brings about large building energy consumption and indoor thermal discomfort at the same time. The poor thermal insulation and high solar penetration are the main reasons. Except for improving the thermal properties of the glazing itself, integrating renewable energy, and especially solar energy, with the glazed facade has also been proven a promising way to achieve building energy saving. Liquid flow window is one of the newly emerged multi-glazing techniques for solar energy utilization and indoor thermal environment regulation.

This book provides systematic information about its basic concept, thermal characteristics, schematic parameters optimization, and application based optimization considering various building types and climates, together with the life cycle impact analysis. More specifically, the thermal performance of liquid flow window with different working fluids is investigated experimentally. And year-round simulations are completed to optimize its schematic design; numerical studies are also carried out to improve its thermal performance for application in different types of buildings and various climates. Life cycle assessment is applied to evaluate its sustainability from an economic, energy, and environmental perspective.

There are totally seven chapters in this book. The first Chapter is introduction of the development of glazing techniques and the liquid flow window. Experimental thermal performance analysis of liquid flow window is in Chapter 2. In Chapter 3, shematic optimization study is included. Thermal performance enhancement for application in two different building types is given in Chapter 4. The application potential analysis in different climates is in Chapter 5. The life cycle payback periods regarding economic, energy and environmental aspects are calculated based on building energy simulation in Chapter 6. Finally, the finished works and the major findings are summarized in Chapter 7.

  This book is a summary of the research works the author has done in liquid flow window thermal performance analysis and optimization. The author appreciates the support of the key Laboratory of Fluid and Power Machinery, Ministry of Education, Xihua University and the Youth Scholar Program of Xihua University. Many thanks are also given to my colleagues and editors who have endeavored to improve the quality of this book.

The author
May 6, 2020

Contents

1 **Introduction** ... 1
   1. 1 Single Glazing ... 3
   1. 2 Double Glazing System ... 5
   1. 3 Multi-glazing System ... 8
   1. 4 Liquid Flow Window ... 10

2 **Experimental Investigation of Thermal Performance of Double Glazing Liquid Flow Window** ... 13
   2. 1 Research Method ... 14
      2. 1. 1 Experiment Set Up ... 14
      2. 1. 2 Measuring Instrument ... 15
   2. 2 Experiments and Results Analysis ... 17
      2. 2. 1 Water Flow Window Performance ... 17
      2. 2. 2 Anti-freezing Liquid Flow Window Performance ... 23
      2. 2. 3 Performance Comparison of Water and Anti-freezing Liquid Flow Window ... 30
   2. 3 Summary ... 31

3 **Schematic Optimization of Liquid Flow Window** ... 32
   3. 1 Research Methodology ... 33
      3. 1. 1 Mathematical Models Development ... 33
      3. 1. 2 Model Validation ... 41
   3. 2 Performance Evaluation with the Variation of Schematic Parameters ... 45
      3. 2. 1 Thermal Performance at Different Cavity Spaces ... 45
      3. 2. 2 Thermal Performance Comparison with Various Pipe-work Designs ... 50
      3. 2. 3 Thermal Performance Evaluation with Variation of Window Dimension ... 57
   3. 3 Summary ... 61

4 Performance Evaluation with PCM Thermal Storage: for Residential and Office Buildings Hot Water Demand Patterns ... 63
4.1 Numerical Models Development ... 65
4.1.1 Mathematical Models for Case 1 ... 66
4.1.2 Mathematical Models for Case 2 ... 67
4.2 Model Validation ... 68
4.3 Numerical Study on Thermal Performance of Liquid Flow Window with PCM Heat Exchanger ... 71
4.3.1 Thermal Performance for Residential Building Application ... 72
4.3.2 Thermal Performance for Office Building Application ... 79
4.3.3 Thermal Performance with Variation of PCM Layer Thickness ... 82
4.4 Summary ... 83

5 Liquid Flow Window Performance under Different Climate Applications ... 85
5.1 Research Methodology ... 86
5.1.1 Model Validation for Anti-freezing Liquid Flow Window ... 86
5.1.2 Numerical Studies in Different Climates ... 89
5.2 Thermal Performance Analysis in Various Climate Applications ... 91
5.2.1 The Effect of Liquid Concentrations on thermal Performance ... 91
5.2.2 Thermal Performance Comparison under Different Climates ... 92
5.2.3 Thermal Performance Improvement in Cold and Extremely Cold Climates ... 96
5.3 Summary ... 98

6 Life Cycle Assessment of Liquid Flow Window ... 100
6.1 The Method of Analysis ... 102
6.2 Life-long Investment, Energy Input and Carbon Emission of Liquid Flow Window ... 104
6.2.1 Investment, Energy Input and Carbon Emission for Window Materials ... 104
6.2.2 Cost, Energy Consumption and Carbon Emission from Construction to Disposal ... 109
6.3 Building Energy Simulation in ESP-r ... 113
6.3.1 Validation Study ... 113

	6.3.2 Building Energy Simulation	115
	6.3.3 Energy Saving Performance of Liquid Flow Window	122
6.4	Payback Period Analysis	125
6.5	Summary	128

7 Summary and Major Conclusions ········ 130
  7.1 Research Work Completed ········ 130
  7.2 Findings and Conclusions ········ 132

References ········ 136

1 Introduction

Climate change and over exploitation of natural resources have been the worldwide environmental issues for decades. One obvious phenomenon in climate change is the rise in global surface temperature of our planet. This is known as global warming. The main cause of global warming is the increase in greenhouse gas emission—a concept first brought out in 1827. Fig. 1.1 shows the global land-ocean temperature change from 1880 to 2014 at magnitude, relative to the 30 years mean temperature from 1951 to 1980. Scientists predict that the global surface temperature may further increase by at least 0.3~1.7℃, and to the extreme, the increase can be even as high as 2.6~4.8℃ if the carbon emission is not well controlled there on.

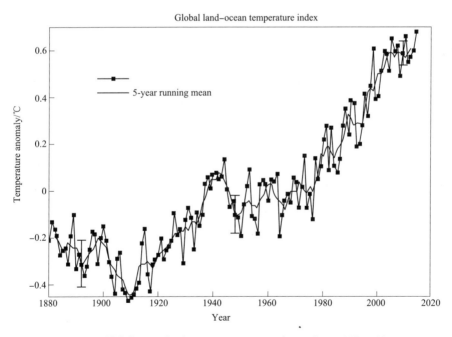

Fig. 1.1　Global mean land-ocean temperature change from 1880 to 2014

The continuous increase in global temperature may lead to great disaster. Melting of snow and glacier has been observed widely in the past years. The increase in sea level can result in flooding along the coast. Habitats of some of the wild animals are threaten and even destroyed, this will break the balance of the ecosystem. Flood and drought caused by climate change have significant effects on agriculture, as well as food and water supply. And this change in climate may even threaten human health directly, especially for the elders and children. To eliminate the threat, the key issues should be found out in the first place, and then alternative approaches for solving the problem can be put forward accordingly.

It has been concluded that the major source of greenhouse gas is carbon dioxide emission, whereas fossil fuel burning in human activity is the main cause of the increase in carbon emission in the past 20 years. Coal contributes to 43% of the total emission; oil, gas and cement are responsible for another 34%, 18% and 4.9% respectively. To solve this problem, there can be many effective ways, such as efficient use of energy, the improvement in forest planting and management, the change of lifestyle and behavior of human beings, and the last but not the least, the exploitation on low-carbon energy resources.

For example, one of the alternative low-carbon energy well considered is nuclear energy. Its environmental impact is much smaller than that of fossil fuel burning. However, the release of radioactive materials and the disastrous damage caused by this has hindered its speed of expansion substantially. Objection from the public is another major factor that affects its widen application. Another clean and safe energy being widely adopted is wind energy. This renewable source has relatively tiny environmental impact. Installation of wind farms has increased greatly and reached 336GW in June 2014. Nevertheless, at the moment the energy generated by wind power contributes to only around 4% of the total electricity consumption worldwide. Its wider application depends strongly on the availability of wind source. Another limitation of wind power application is the large space requirement for the wind farm.

Solar energy has much greater application potential when compared to other traditional or clean resources. The utilization of solar energy has been widely studied for centuries. Many active and passive technologies have been proposed and well developed in the industry. Further research works with regard to their performance enhancement have been carried out by researchers all around the world. The most widely accepted technologies are the photovoltaic system and the solar thermal system.

The application of solar collectors in buildings can release the environmental burden effectively, since the building sector contributes to a large portion of the overall energy consumption. This contribution ranges from 20% to 40% in developed countries. The residential and commercial buildings in Europe account for more than 40% of the overall energy consumption. In China, the building contributes to about 40% of the total energy consumption. Of the energy consumed by buildings, a large portion is used in the air-conditioning system for space cooling purpose. The increase in global air temperature will further intensify this consumption because of the increase in ambient temperature. Thus, one of the effective approaches to solve the environmental problem is to make better use of renewable energy and construct more energy efficient buildings, which will have less energy demands and environmental impacts, because of their better use of natural resources and creative passive technologies.

How to make good use of solar energy in well-designed buildings has been widely studied. The proper arrangement of building orientation is the simplest ways to achieve this goal, and an alternative approach is to improve the design of building envelope. Good envelope design can limit its surface area exposed to the outdoor environment, and well control the solar penetration and the space heating/cooling loads consequently. Building elements like foundation, roof, walls, doors and windows are generally taken as building envelope components. Materials of the building envelope components also have dominant effects on heat flow between the indoor and outdoor spaces. The overall heat transfer coefficient varies greatly with the composite layers of building materials. The energy consumption reduction may reach 40% by adopting well designed envelope, according to the buildings in previous studies in China and Greece. Passive solar walls, lightweight concrete walls, ventilated walls and latent heat storage walls have been used in the construction industry. Compared to room heat gain brought by opaque walls, the heat transfer either inward or outward through window systems can be even larger. As a matter of fact, window plays an important role in building by providing natural light and creating a comfortable visual environment for occupants. Window system with heat transfer coefficient as small in magnitude as possible is more favorable for energy efficient building design when considering the energy consumption of the air-conditioning system. However, a good visual effect and the readiness for access to natural outside view are also desirable. Nowadays, more and more highly glazed buildings are built, thus the thermal blocking performance of the building envelope are further weakened. This makes it necessary and urgent to improve the thermal performance of window material and glazing products. Lots of advanced single and multi-glazing systems have been developed in the past years to meet the above requirements.

1.1 Single Glazing

Clear glass is the most basic one with glass floated over a bed of molten tin. It has high solar transmittance, and so affects the thermal and visual comfort of occupants in hot summer and cold winter seasons. Many alternative glazed products were developed to deal with these problems.

Some of the commercialized products include tinted, refelctive/anti-reflective and dynamic glazing. Tinted glass has the largest solar absorptance in certain spectrum range, so it is also called as 'absorptive' glazing. Large amount of solar thermal energy is stored in the glass pane and its surface temperature is higher than the other glazing systems. The indoor cooling load can be effectively reduced if the absorptive glazing is used in combination with low-E glazing. To reduce the solar thermal transmittance, reflective glazing is also developed. With this system, the indoor heat gain can be reduced in larger degree compared to the tinted glass, and the visible light transmittance is not deteriorated. Mohelnikova compared the performance of different materials used as reflective coatings. By using both numerical and experimental methods, it was hoped to make some improvements to window design by examining the optical and thermal properties of glasses. The results indicated that double-metal glazing is better than single one. Arsenault et al. studied the response of people to visual condition under different glazing color types. The research was carried out experimentally for investigating daylight quality, arousal, and switch-on patterns of artificial lights. They found that a warm color like bronze is more popular than cold colors.

On the contrary, anti-reflective glazing has been proposed to increase the solar heat gain through window. This can be combined with other glazing systems for practical use. Andreas et al. investigated the visual and energy performance of different windows (clear floating, low-E glass and electrochromic) with anti-reflection coatings. Both solar transmittance and visible transmittance were found improved by the addition of anti-reflection coating. The work of Rosencrantz et al. gave very similar results. They pointed out that the improvement of visual and thermal transmittance could be achieved by adding an anti-reflective coating to a double glazing low-E window system.

Smart window (electrochromic window) is another novel product which can change solar factor depends on the outdoor and indoor conditions. Through this, the energy consumption for cooling and heating can be reduced effectively. The optical properties of the electrochromic film vary in a reversible and persistent way under the action of voltage pulse; it can be switched between different states, including transparent, absorbing and reflecting states. Besides, there are also other materials that can be used for smart glazing. The performance of electrochromic, dispersed liquid crystal and dispersed particle glazing were compared in the study of Lampert. The electrochromic glazing was found more suitable for buildings with an emphasis on the comfort and well-being of users. Similarly, Piccolo et al. found that a reduction of discomfort glare could be achieved by using electrochromic glazing windows. Visual and energy management of smart (EC) window were also investigated by Gugliermetti et al. using the IENUS simulation package. Temperature variation of smart window under direct solar radiation was studied and the results were compared with that of double-glazing glass. The results showed that the temperature increase of smart window appears to be higher. In addition, to meet different requirements of occupants, Jonsson et al investigated the control methods for different combinations of smart windows.

A self-cleaning glazing was also developed to provide a positive environmental impact. There is a photocatalytic reaction in the thin coating on the glass, such that the water carries dirt off when it falls on the glass. This product can help to reduce both cost and energy used during the cleaning process. Midtdal et al. reviewed the self-cleaning products available on the market and compared their performance by measuring the cleaning effect of different products. They concluded that different self-cleaning products should be adopted according to the conditions they were used. The single glazing technologies like low-E, smart windows, solar cell et al., can also be combined to form double or multi glazing systems with better thermal performance as given below.

1.2 Double Glazing System

Low-E glazing refers to low emissivity glazing, which is of simple structure and widely used in industry. Its property varies according to the wavelength of the incident solar radiation. The Low-E coatings are usually applied to the second or third surface of a double glazed unit, which is known as Insulated Glazing Unit (IGU), for the purpose of protection it from deterioration. The coating on the third surface allows the short wave solar radiation to pass through and the long wave thermal radiation from the inside environment is selectively reflected back into the building. Low-E glazing can be divided into two categories: soft and hard glazing. They have different optical properties because their different ways of manufacture. Under certain circumstances, a higher visible transmittance is required. Hammarberg et al. found out a way to solve this problem by depositing thin films of SiO_2 on both sides of the low-E based coating. In their study, the visible transmittance can be increased by 9.85% to 91.5%. Because of its good performance and flexible properties, it is also considered to be used in building retrofitting works. The users' requirements should be considered in this process and the cost can be reduced greatly by choosing the proper contractor.

Similar to IGU, the air or gas filled units were also developed. Thermal performance of air filled window is affected by the gap dimension. Furthermore, its thermal performance can be modified by filling different gas materials, like absorbing gas or inert gas. In this type of window, the proper selection of glazing material is also of great importance in improving the energy saving performance. And the edge effect caused by the frame and spacer material should also be taken into consideration. The adoption of aluminum spacer increases the risk of surface condensation. And Low-E glazing was one of the methods adopted to avoid this condensation problem. Three different types of spacer materials were tested in the study of Song et al. Performance of the thermally broken aluminum spacer was found better than the traditional aluminum spacer for preventing condensation, and the thick-walled plastic spacer was the best among the three test cases.

Vacuum window is another design which adopts an evacuated space to eliminate conduction and convection. Thus its thermal insulation performance is better than the traditional double glazing product, while comparable natural lighting can be obtained. Fig. 1.2 shows the structure of one vacuum window, the evacuated gap is formed and supported by small pillars at equal space and dimension. By incorporating a low-E coating on one of the inner surfaces, the radiative heat transport was found further eliminated, and the energy saving performance could be enhanced.

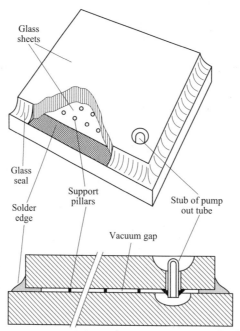

Fig. 1.2 Schematic structure of vacuum window

The concept of Phase Change Material (PCM) filled window was also proposed to reduce the indoor cooling load. PCM is filled in the cavity between two glass panes to maintain a more stable room temperature, and consequently better thermal comfort in room environment. The performance of PCM filled window was compared to other products to study its potential for application in hot climate. Its optical properties were found vary in different patterns with the wavelength, and affected by the thickness and properties of PCM. However, the visible transmittance of this kind of window is not large enough to meet users' requirements. The visible transmittance changes from 90% to 40% in liquid and solid state, respectively. Appearance of double glazed window with clear glass and PCM filled in the cavity is shown in Fig. 1.3; in which the PCM is in the solid state. The change of its appearance during the day may limit its application. But it may have the chance to be used in conditions when glare should be prevented. For the selection of PCM to be used in window cavity, it should be determined by the weather condition. The PCM with higher melting temperature is recommended for hot climate. In cold climate, however, the melting temperature of PCM should be low to improve the thermal comfort level of users.

Fig. 1.3 Appearance of double layer window with clear glass and PCM in cavity

Ventilated window is another novel design that uses the flowing air to discharge the thermal energy of a window system. Zhou et al. reviewed the application of double-glazed ventilating facade on buildings in the hot summer and cold winter region in China in order to meet the requirement of sustainable development. All the research results showed that it had great potential for use in China if the system was properly designed. And blinds integration is one of the options. Heat recovery ability of double ventilation window in cold climate was also studied by Carlos et al. Chow et al. studied the application potential of ventilated window in warm climate like Hong Kong and found that the reversible ventilated window, which has different operation modes during summer (with outdoor airflow in the ventilated gap) and winter (with room airflow in the ventilated gap) seasons, was not suitable for areas with short or no winter periods. Gugliermetti et al. investigated the energy saving potential of the fully reversible window and found its advantage was especially evident in area with distinct cooling and warming climate. More researches were carried out with experiment and simulation tools regarding ventilated window systems, the results indicated that PV ventilated window performed better compared to single absorptive and single PV window in both tropical and template climates.

1.3 Multi-glazing System

Multi-glazing system has improved thermal insulation performance because of the increased gap space and window thickness. Similar to double glazing system, the gap filling material and glazing material selections are the determinant factors for its overall thermal performance. The mostly commonly adopted are evacuated, PCM filling, air flow or gas filling.

The benefit of the vacuum gap is conduction and convection elimination. However, thermal conduction of edge seal should be taken into cosideration, especially in windows with small dimension. Since this will cause the increase of U-value in the area nearby. The use of frame material with low conductivity, like wood, is one of the options to solve this problem. Besides, the vacuum glazing is under the risk of breakage under even small load because of the pressure difference between the inner and outer sufaces. In the study of Fang, tempered glass, which is four to ten times stronger than the annealed glass, was proposed and the corresponding thermal performance as compared to the window with normal annealed glass was studied. As a result, overall U-value of the window was reduced because of the reduction of support pillar numbers. Two connected air chambers were formed to realize the pressure balance between the two surfaces of the evacuated glazing in the study of Bao as given in Fig. 1.4. Thermal performance of the vacuum glazing is sensible to the selection of glazing material, and low-E glazing was proven a promising choice to reduce heat transfer coefficient of the window. In triple glazing, it is suggested to place one low-E coating in each of the evacuated gap.

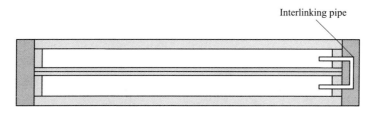

Fig. 1.4 Hybrid pressure balance vacuum glazing unit

In double glazing PCM window, extra room heat gain appears during the summer night time and after the through melting of PCM. Thus, triple PCM window was adopted to eliminate the overheating problem. By adding an air layer, the inner surface temperature of window can be lowered and a more stable temperature can be maintained in the summer season; while in the winter season, a higher inner surface temperature is realized. Then both cooling and heating energy consumptions can be reduced. Its energy performance is also superior to the triple glazing window with two air filling chambers. Similarly, the window configuration, like the proper position of the PCM chamber is significant for improving thermal performance of the window. The PCM layer was suggested to be placed near the outdoor environment for subtropical application. Besides, the thickness of the PCM layer is also a determinant factor that affects the energy saving performance of multi-layer PCM window, and a thickness not more than 20mm was recommended in the study of Li.

A switchable exhaust air multi layer window given in Fig. 1.5 was put forward by Zhang et al. The treated exhaust air from the room environment flows to the outdoor environment through the switchable air chamber, which is the outer chamber in summer operation and inner chamber in winter operation. By doing this, temperature difference between the inner glazing surface and the room environment can be reduced. Results indicated that thermal transmission through the window could be lowered effectively in both peak summer and winter days as compared to normal double and triple glazing system. Except for the single switchable flow path, the ventilated air can also flow in two independent flow paths. This type of window was proven effective in reducing heating load, while the contribution to cooling energy saving was less significant. Thus, it is suitable for cold erea with heating demand. Following this, a triple layer window with air flows in the two series connected flow paths (shown in Fig. 1.6) was proposed, the flowing air was heated by the recovered thermal energy and the incident solar thermal energy. Thus a larger temperature increase can be realized. Its thermal performance for application in moderate climate was studied. Very promising heating energy saving was obtained, while thermal comfort of occupants might be affected because of the cold wall effect caused by the supply air, especially under the extremely cold climate.

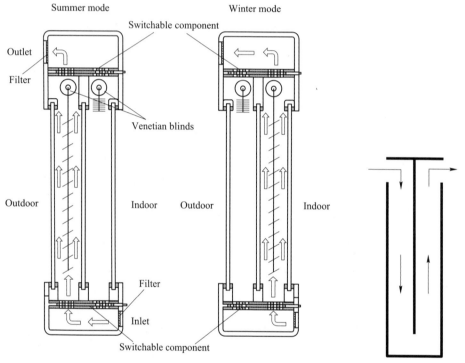

Fig. 1.5　Schematic structure of switchable exhaust air triple glazing window

Fig. 1.6　Series connected double path window

1.4　Liquid Flow Window

Different from the previous techniques, liquid with better thermal properties is used as the circulating medium in liquid flow window instead of the commonly usd gas material. Fig. 1.7 shows the structure of liquid flow window carrying a heat exchanger. In this design, the liquid stream circulates in a closed flow path formed by the window cavity, the heat exchanger and the connection pipe under the buoyance driven force. And thermal energy carried by the warm water stream is released to the cold feed water at lower temperature in the heat exchanger. Thermal capacity of liquid is much larger than that of gas material and thus larger amount of thermal energy can be extracted by the flowing liquid. Furthermore, thermal energy collected by the liquid medium can be taken as low grade heat source for building use.

Thermal characteristics of water flow window in cooling dominated climate have been widely studied. Under such application, it contributes to building energy saving from two aspects: (1) cooling energy saving, and (2) water preheating. By using water flow window, room temperature in the summer season was found lowered by 18℃ for a non-air-conditioned room in Spanish. Thermal performance of water flow window was found affected by the water flow velocity, solar intensity, glazing properties as well as the schematic design. The effect of water flow velocity is more significant when the flow velocity is smaller than 0.01m/s. And thermal efficiency can be regulated by changing the velocity because of the variation in g-value. With absorptive glazing, larger amount of thermal energy can be extracted by the water layer, and room heat gains through water flow window can be reduced by 52% and 32% as comparing to single and double glazing systems. The reductions are lowered when anti-reflective glazing is in use; they are 35% and 22% respectively. Based on the schematic optimization study, the proper dimension parameters and the water layer thickness were recommended in the study of Lyu. And pipe-work simplification was also found useful for thermal performance improvement. To cater for different hot water demand patterns of residential and commercial buildings, PCM was integrated for thermal storage.

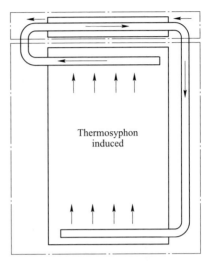

Fig. 1.7 Front view of the buoyance driven liquid flow glazing

For application in area with cold winter and low ambient temperature, both passive and active approaches are available. The passive approach is to lower the freezing temperature of the liquid medium by adding anti-freeze to the water layer. By doing this, promising thermal performance can be maintained even in extremely cold climate. And temperature controlled water supply is the active approach; the temperature regulation could be realized by combing with geothermal or solar thermal systems. By supplying warm water in the heating season, room heat loss could be reduced and thermal comfort of room environment was enhanced. Warm water was supplied to the embedded water pipe instead of the window cavity in the study of Shen et al. , room temperature was found elevated effectively and the heating load was reduced. The heating load reduction is related closely to the warm water temperature, which should be determined considering the local weather characteristics. Zero heat loss through the window was achievable by supplying warm water at proper temperature. Besides, by supplying warm water to the building facade, more stable room temperature and better thermal comfort could be maintained.

To further improve its thermal performance and realize different functions of room cooling, heating and warm water generation, multi-layer liquid flow window were developed. In the study of Li, thermal performance of a double channel water flow window was investigated, the outer chamber was used for thermal collection, and the inner chamber was used for room thermal environment regulation. Low-grade geothermal energy could be used to generate cold and warm water in summer and winter seasons. In a fluidized glazing, the two liquid chambers were seperated by an IGU. The use of IGU was to enhance thermal insulation. Heat transfer between the indoor and outdoor environments was thus further reduced. In the study of Villasante, it was found that thermal performance of a triple glazing liquid flow window could be further regulated by adjusting the thickness of the liquid layer.

2 Experimental Investigation of Thermal Performance of Double Glazing Liquid Flow Window

With the appearance of energy and environmental crisis, the awareness on building energy saving performance improvement and sustainable building techniques development are becoming higher and higher. Thus, thermal performance assessment regarding the new emerging techniques is widely conducted. The multi-glazed liquid flow window is a novel energy efficient window/façade technique for green and sustainable building development. To explore its effectiveness and advantages in building energy saving application, thermal performance analysis should be conducted.

Experimental testing is one of the most popular approaches for thermal performance analysis. It is capable of evaluating thermal performance of glazing and window techniques in a wide range from energy saving performance to indoor thermal environment regulation, as well as visual and thermal comfort analysis. For example, the solar transmission through various kinds of window materials were successfully investigated with combined experimental and theoretical approaches in the study of Kumar et al. The cooling and heating energy saving of spectrally selective coated glazing comparing to low-E coating and normal glazing was studied, and energy savings of 21% ~ 23% were obtained. The experimental mock-up cell was created and the testing method for human response evaluation to luminous environment was proven economic and suitable for other smart window systems. The risk of overheating and over illumination was investigated under various weather conditions in the study of Ajaji et al.

The experimental method is widely applied to thermal performance analysis of various types of window products. The window can be static, like simple single glazing or multi-glazing with different filling materials. For window with dynamic properties, the same method can be used. Thermal and optical performance evaluation of dynamic thermochromic and electrochromic glazing were completed successfully. Energy saving performance evaluation of a triple glazing window making use of low-grade exhaust air, which has similar structure with the current liquid filled window, was completed. Comparing to the traditional triple glazing window, the annual accumulated cooling and heating load reductions were found as large as 25.3% ~ 50.1%. By replacing the flowing air stream with liquid medium, which has larger heat capacity, thermal transportation efficiency should be further improved. Thus, energy saving potential of the current multi-glazed liquid filled glazing should be very promising.

In this chapter, thermal performance of double glazing liquid flow window will be assessed experimentally. Window surface temperatures, thermal transmission through the window, which affects building cooling and heating energy consumptions, and useful water heat gains will be compared to evaluate its impact on indoor thermal environment and building energy saving. The experimental results are also widely used for numerical and analytical method validation because of its good accuracy and reliability. It is so for the current study, the obtained experimental results in this chapter will also be used for numerical analysis validation in the following chapters.

2.1 Research Method

2.1.1 Experiment Set Up

A prototype of liquid flow window with glazing dimensions of 1.2m(H) × 0.8m(W) × 0.012m(D), including a 20mm cavity was constructed. The double pipe heat exchanger was of optimized diameters of 30mm and 20mm for the outer and inner pipes, respectively. This counter flow heat exchanger was connected to the glazing system by flange joint as shown in Fig. 2.1, for the convenience of thermocouple insertion. Another two sets of flanges were used for the connection between the cold water main and the heat exchanger, for the same reason. There were two extra pipes at the sides of the heat exchanger setting as an air vents and to cater for liquid expansion. Two pieces of sight glass were installed at these pipes for observing the change of liquid level during the experimental tests. On the two headers there were ϕ2mm openings at equal spacing of 30mm.

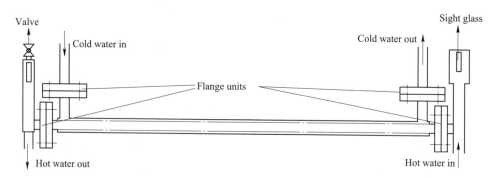

Fig. 2.1 Falange connection of window with heat exchanger

The experiment was carried out in Changzhou, Jiangsu province from late September to early October. The test cell shown in Fig. 2.2 was on the roof top of a five-story building on the campus of Changzhou University. Its overall dimensions were 3m(L)× 3m(W)× 3m(H). The window system was installed south facing to receive the maximum amount of solar energy. The selected glazing type was absorptive glass panes; this contributed to better thermal absorption and water pressure resistance. The thermal and optical properties of the absorptive glazing used in the experiment are listed in Table 2.1, with the same reflectance at front and back sides. The experiment was completed with different working fluids of water and anti-freezing liquid in the window cavity. Cold water in the open circuit was connected to the water main, and it was drained away after heated by the hot fluid in the close circuit in the experiment. This was expected to be delivered to hot water system in practical application.

Fig. 2.2 Test cell of liquid flow window

Table 2.1 Optical and thermal properties of 12mm absorptive glazing used in the experiment

Thickness/mm	Visible transmittance	Visible reflectivity	Solar transmittance	Solar reflectivity
12	0.61	0.06	0.29	0.045

2.1.2 Measuring Instrument

Global solar radiation on the horizontal surface was measured with Pyrometers as shown in Fig. 2.3. Two CMP6 pyranometers manufactured by Kipp & Zonen were used for global solar radiation recording. Its sensitivity was 19.76μV/(W · m^2) and the operational temperature range was from $-40\,^\circ\mathrm{C}$ to $80\,^\circ\mathrm{C}$. Its directional error was less than $20\mathrm{W/m^2}$. And it was also used as input in the validation test of the numerical method.

Fig. 2.3 Pyrometers used for measuring of solar radiation

Glazing surface temperatures, inlet and outlet temperatures of both hot and cold fluids in the heat exchanger, and air temperatures of the indoor and outdoor environments were measured with type T thermocouples. They were suitable for temperature measurements in the range from $-200\,^\circ\mathrm{C}$ to $260\,^\circ\mathrm{C}$ with an accuracy of $\pm 1\,^\circ\mathrm{C}$. The temperatures of outer glazing were detected at three different heights, including the lower, middle and upper levels. Two thermocouples located at the left and right sides were installed at the upper and lower levels of outer glazing, and the average values of the two measuring points at the same height were used in performance evaluation. This is because temperature increase along the height of the window is quite stable and temperatures at the left and right corners are very close. For the inner glazing, only three thermocouples were installed at three different heights which are the same as those of outer glazing. The temperatures at the lower level, the upper level and the center point were named as T_1, T_2 and T_3, the distribution of measuring points are as shown in Fig. 2.4.

For the water temperature measurement in the heat exchanger, thermocouples were put into the water stream directly through the flange connection point, and two thermocouples were installed at each point to acquire more accurate measurement. The exception was the one at the outlet of the hot water stream, which was adhered to the outer surface of the hot water pipe. This temperature data was only used as a reference point to compare the difference in temperatures measured at the stream and at the outside surface of the pipe. The average temperature of these two detectors at each measuring point was used in the energy flow calculation. Another two thermocouples were used to measure the ambient temperature outside the test cell. The air temperature and cold feed water inlet temperature

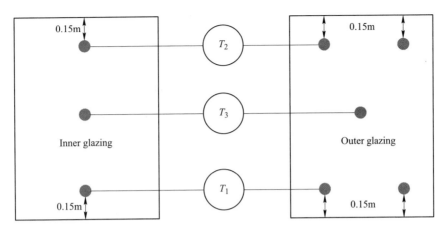

Fig. 2.4 Distribution of temperature detection points at window surfaces

were used as input in the validation test, in which the measured glazing surface temperatures were compared with the simulation results obtained from the self-developed computer program.

The liquid-flow rate or velocity in the window cavity could not be measured directly because of the difficulty in laboratory set-up. But the measurement of the cold feed water flow rate was convenient, and the data could be used in subsequent energy calculations and model validation tests. The flow rate was measured in term of both mass and volume flow rates, with the use of measuring cup, stop watch and electronic scale. The comparison between the two measurement approaches showed good agreement, which justified the method in use.

All the acquired data, including temperatures and horizontal solar radiation level, were recorded by the Agilent 34970A data logger in every two minutes, and the solar radiation level was recorded as DC current signal. They could be stored in and downloaded from a desktop computer.

2.2 Experiments and Results Analysis

2.2.1 Water Flow Window Performance

The experiment with water as working fluid was carried out in three consecutive days from $25^{th} \sim 27^{th}$ September, and they were taken as Day 1 to Day 3 in the results comparison and analysis. Experimental testing was conducted during only the daytime, from 7am to 6pm. The effective temperature increase appeared between 10am and 4pm and with fairly weak solar radiation at 5pm. The visual effect of water flow window is shown in Fig. 2. 5, in which the position of arrow is the interface of water and air. Good visual performance can be observed for the view from inside to the outdoor environment. This is because this product has a high visible transmittance of 0. 61 (Shown in Table 2. 1).

Fig. 2. 5 Visual effects of water flow window at late afternoon

The recorded solar radiation levels on horizontal surface are given in Table 2. 2,

Table 2. 2 Hourly average solar radiation condition during the 3 test days (10:00~17:00)

Hour of day	Incoming global solar radiation/W · m^{-2}		
	Horizontal		
	Day 1	Day 2	Day 3
10:00	722. 68	496. 69	575. 11
11:00	617. 05	333. 69	617. 31
12:00	409. 33	408. 15	629. 59
13:00	393. 41	413. 55	649. 20
14:00	286. 69	508. 20	579. 34
15:00	206. 28	229. 04	138. 94
16:00	111. 86	90. 93	46. 98
17:00	27. 40	27. 04	19. 34
Average	346. 84	313. 41	406. 97

whereas the measured ambient and cold feed water temperatures are in Table 2.3. The solar radiation level was higher on the Day 3 as compared to the previous two days. The highest level of the computer record was 722.68W/m^2, which occurred at around 11am on Day 1. The peak values occurred at around 2pm and 1pm respectively for the following two days.

Table 2.3 Measured ambient temperature and cold water inlet temperature at the daytime during the 3 test days (10:00~17:00)

Hour of doy	Temperature/℃					
	Outdoor environment			Cold water inlet		
	Day 1	Day 2	Day 3	Day 1	Day 2	Day 3
10:00	29.49	27.01	30.04	30.77	28.75	30.39
11:00	29.92	27.01	31.04	31.98	27.55	32.32
12:00	28.65	28.01	31.92	30.11	28.70	33.39
13:00	28.66	28.74	32.56	29.62	29.75	34.65
14:00	28.01	30.11	32.35	28.80	31.02	34.35
15:00	27.49	27.90	30.57	28.51	28.91	31.48
16:00	26.65	27.17	29.55	27.11	27.56	29.69
17:00	25.65	26.45	28.86	25.82	26.48	28.68
Average	28.07	27.80	30.86	29.09	28.59	31.87

The highest ambient and cold water inlet temperatures appeared at almost the same time with the strongest solar radiation. The ambient temperature on Day 3 was higher than the other two days, and the cold water inlet temperature was higher than the ambient temperature through out the three days. This was mainly caused by the exposure of the supply water pipe under direct solar radiation.

The measured cold water flow rates are listed in Table 2.4. This is found closely linked to the system thermal efficiency because the effectiveness of the heat exchanger is influenced by the mass flow rates of the two fluid streams. The measured flow rates were subsequently used as input parameters in the model validation test described in the following chapters.

Table 2.4 Measured cold water flow rate during the 3 test days (10:00~17:00)

Time of day		10	11	12	13	14	15	16	17
Flow rate /mL · min^{-1}	Day 1	180	150	140	170	170	160	180	170
	Day 2	200	180	180	180	180	165	170	170
	Day 3	160	180	180	180	165	140	160	160

There exists time delay for the heat transfer from the outer glazing to the water layer. The outer glazing was always heated in the first place, and then the thermal energy was transferred inward to the water layer and the inner glazing by convection and conduction. The temperature variations at the hot water inlet and at position T_2 of the outer glazing during the three consecutive days are shown in Fig. 2.6. The graphs indicate that the peak value of hot water inlet temperature always occurred later than that at the upper position 'T_2' of the outer glazing. It can be seen that the hot water inlet temperature was lower than the temperature T_2 at the upper level of the outer glazing in the early morning. This was later higher than T_2 of the outer glazing after reaching the peak point as a result of the delay in heat transfer. There was an exception at 11am of Day 2 that the hot water inlet temperature was higher than that at the upper surface of the outer glazing. This was caused by the strong solar radiation at 10am, and was found 496.69W/m². The solar radiation level in the following 3h was then decreasing, thus it was the high outer glazing temperature at 10am that caused the water temperature increase at 11am.

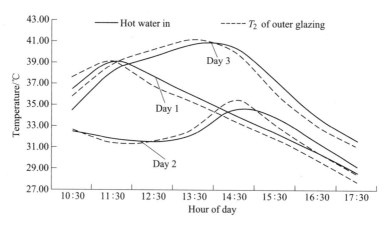

Fig. 2.6 Variation of hot water inlet and T_2 of outer glazing for the 3 test days from 10am to 6pm

This phenomenon was also true for the variation of outer and inner glazing surface temperatures. The temperature of the outer glazing increased within a short time because this glazing was exposed to the direct solar radiation. Thus, the temperature at the center of the glass (T_3) was closer to the upper surface temperature (T_2) as shown in Fig. 2.7. Besides, the difference between T_2 and T_3 was smaller under stronger solar radiation. This could be observable for all the three days, ie. at 10am on the first day, 10am and 14pm on Day 2 and Day 3. However, for inner glazing, the temperature at the center point (T_3) was closer to the lower surface temperature (T_1). This was caused by the delay in heat transfer and the thermal energy extraction of the heat transfer fluid, which lead to the reduction in thermal absorption of the inner glazing. The variation of inner glazing surface temperature at the three measuring points is shown in Fig. 2.8. Different from the outer glazing case, a stronger solar radiation leads to a smaller difference between T_1 and T_3 (temperature at the center point is closer to that of bottom level). In this case, the convection heat transfer was enhanced under the strong solar radiation and thus the amount of thermal energy taken away by the water layer was increased, and accordingly the rate of temperature increase of the inner glazing was reduced.

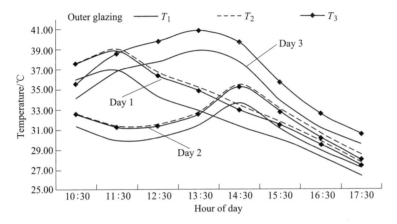

Fig. 2.7 Surface temperature of outer glazing at different heights for the 3 test days from 10am to 6pm

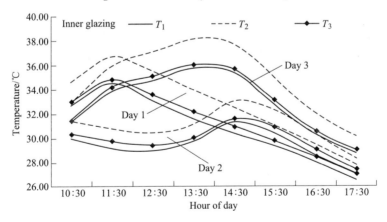

Fig. 2.8 Surface temperature of inner glazing at different height for the 3 test days from 10am to 6pm

The variation of hot and cold fluid temperatures in the heat exchanger is shown in Fig. 2.9 and Fig. 2.10. The temperature difference between the outlet of hot and cold water streams was large when the solar radiation was strong, and the temperature increase of cold feed water was also large at this moment because of the improved heat transfer coefficient. But this difference became smaller and smaller, and the temperature increase of the cold water was also reduced as solar radiation level reducing. This was because the hot fluid flow slowed down when the solar radiation was at the extremely low level. Under this circumstance, the outlet temperature of cold water was almost equal to that of hot water in the late afternoon, and the thermal energy carried away by the hot water stream could be fully released to cold water. This was observed for all the three days. The maximum difference between the hot and cold water outlets was 1.27℃ on Day 1, and this was always smaller than 1℃ for the following two days.

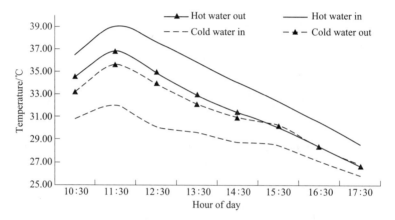

Fig. 2.9 Variation of water temperature in the heat exchanger on the first test day from 10am to 6pm

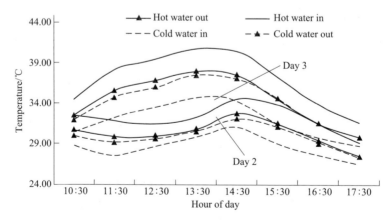

Fig. 2.10 Variation of water temperature in the heat exchanger on the last two test days from 10am to 6pm

The main contributions of this innovative system for application in the cooling dominating climate regions are two folds: the reduction in room cooling load and the useful water heat gain. The daily water heat gain of this window system and the daily thermal efficiency are as listed in Table 2.5. The latter was determined based on total water heat gain and the corresponding incident solar radiation from 10am to 6pm. The magnitude is influenced not only by the solar radiation level, but also by the cold water heat gain.

Table 2.5 Daily water heat gain and thermal efficiency during the 3 test days

Date	Day 1	Day 2	Day 3
Water heat gain/J	207.31	126.99	205.81
Thermal efficiency/%	9.41	6.66	8.01

The solar radiation level on Day 2 was only 9.6% lower than that on Day 1, but the thermal efficiency was about 29.2% lower. This was caused by the high cold water inlet temperature which was closely related to the ambient temperature and the solar radiation condition. The daily average temperature of Day 2 was almost equal to that of Day 1 though the solar radiation level was lower. The hot water inlet temperature of Day 1 was much higher than that of Day 2; and the difference was larger than 2℃. The temperature difference between the hot and cold water inlets were 5.2℃ and 3.52℃ on the first and the second test days respectively. This was another factor leading to the larger cold water heat gain on Day 1. The thermal efficiency of Day 3 was smaller than that of Day 1. This was also caused by its higher cold water inlet temperature. The daily average difference between the hot and cold water inlet temperatures on Day 1 was 5.12℃. The magnitude was close to the corresponding difference on Day 1. However, this difference of Day 1 was about 1.1℃ higher than that of Day 3 considering the period from 10am to 1pm when the solar radiation was strong. This was caused by the highest cold feed water inlet temperature during this period since the supply cold water was heated to a greater extend in the supply pipe under the stronger solar radiation condition of Day 3. The small temperature difference during this period limited the improvement in the overall thermal efficiency. Thus the system thermal efficiency could be further improved by adding better insulation onto the cold water supply pipe. However, it should also be noticed that exposure of supply pipe to sunshine is good for solar energy utilization because water can be preheated before delivering to the hot water system. This is especially true in the summer season.

2.2.2 Anti-freezing Liquid Flow Window Performance

Anti-freezing liquid has been widely used in solar water heater system in cold countries to prevent freezing of working fluid. This is for maintaining normal system performance and protecting equipment from damage during winter operation. In the present study, it was also adopted for anti-freezing purpose though transparent insulation could also protect water from freezing. But transparent insulation was not suitable for this system application considering the following viewpoints. Firstly, it could not be used in extremely cold climate. Secondly, the visual comfort of occupants would be affected negatively. And lastly, the increase in system weight would have additional demands on the whole building structure. However, with anti-freezing fluid, the freezing point of water could be decreased by certain degree depends on the percentage of anti-freezing material in the working liquid. The fluid used in this study was solar heat transfer fluid produced by the Suntask Company. This is in blue color. The replacement of water with the solar heat transfer fluid may affect the visual performance of the system, and pictures of partly filled window with anti-freezing liquid taken in the early morning and late afternoon are shown in Fig. 2.11. The positions of arrow show the interface of water and air. In the early morning, the visual effect of the liquid filled part was even better. This was owing to the presence of water vapor in the upper part.

Fig. 2.11 Visual effect of anti-freezing liquid filled window viewing from inside to outside
(a) Early morning; (b) Late afternoon

This solar heat transfer fluid used in the experiment was a mixture of distilled water and propylene glycol at a ratio of about 2 : 3. This means that 40% of propylene glycol was mixed with 60% of distilled water by volume. This could help to reduce the freezing point of the liquid mixture down to −15℃. And it is also adequate to cater for the winter operation in Changzhou, as well as in most other cities in the middle part of China. Some of the thermo-physical properties of the liquid used in the experiment are listed in Table 2.6.

Table 2.6 Thermal properties of anti-freezing fluid used in the experiment

Freezing point/℃	Working temperature/℃	Expansion coefficient	Conductivity /W·(m·K)$^{-1}$	Heat capacity /kJ·(kg·K)$^{-1}$
−15	−15~130	4.0~5.8	0.39~0.42	3.4~3.8

Success of the outdoor test relied highly on the available solar radiation during the experiment. And thus, this second experiment was first carried out on 28th Sep., and then on 2nd and 3rd Oct. because of the bad weather condition in between. These three test days were taken as Day 4, Day 5 and Day 6. Similar to the first experiment, the weather parameters including the solar radiation level and air temperatures were recorded and used in performance evaluation as well as in mathematical model validation. The solar radiation incident on the horizontal and vertical surfaces is listed in Table 2.7. The solar radiation level during these three days was higher than the previous three days when the experiment of water flow window was carried out, especially on Day 4.

Table 2.7 Hourly average solar radiation condition from Day 4 to Day 6 (10:00~17:00)

Hour of day	Solar radiation/W·m^{-2}		
	Horizontal		
	Day 4	Day 5	Day 6
10:00	584.08	715.42	657.24
11:00	722.00	779.93	637.73
12:00	738.08	713.00	614.12
13:00	662.81	491.39	579.95
14:00	523.78	300.86	510.54
15:00	345.94	268.95	316.39
16:00	152.80	50.71	138.43
17:00	33.99	10.20	28.52
Average	470.43	416.31	435.36

The outdoor air and cold water inlet temperatures are listed in Table 2.8. Similar to the results of water flow window, the cold water inlet temperature was higher than that of the air temperature because of the weak thermal insulation applied on the feed water pipe. The difference was particularly large on Day 4, when the cold feed water inlet temperature was about 3℃ higher than the air temperature from noon to 2pm. The daily average increase from air temperature to cold feed water temperature was about 2℃. This was much higher than the temperature increases during the other days and that in the water flow window experiment. It happened under extremely strong solar radiation, the feed water in the naked supply pipe was heated by a large extent. The results were similar to those of Day 3; in that a stronger solar radiation caused a larger difference between the ambient temperature and the cold water inlet temperature. Thus the temperature difference between the hot fluid and cold water inlets would be small.

Table 2.8 Measured ambient temperature and cold water inlet temperature from Day 4 to Day 6(10:00~17:00)

Hour of day	Temperature/℃					
	Outdoor environment			Cold water inlet		
	Day 4	Day 5	Day 6	Day 4	Day 5	Day 6
10:00	28.88	27.96	27.45	30.45	28.07	27.06
11:00	30.59	28.74	27.55	33.21	28.19	27.73
12:00	31.14	29.71	27.85	34.27	30.29	28.28
13:00	31.87	27.56	28.35	34.97	28.70	29.02
14:00	31.52	26.62	28.30	34.26	27.15	29.64
15:00	30.72	26.36	26.64	32.00	27.23	28.65
16:00	29.78	24.63	24.75	30.35	25.44	26.79
17:00	25.01	23.89	22.93	25.38	24.32	24.96
Average	29.94	26.93	26.73	31.86	27.42	27.77

The hot water inlet temperature and the corresponding difference with cold feed water inlet temperature are listed in Table 2.9. The temperature difference between the hot and cold water inlets was as small as 4.65℃ on Day 4. This was much smaller than the other two days that carried the values of 7.2℃ and 8.48℃. And thus the useful water heat gain would be affected greatly.

Table 2.9 Measured hot fluid inlet temperature and temperature difference between hot and cold fluids inlet from Day 4 to Day 6(10:00~17:00)

Hour of day	Hot water inlet temperature/℃			Temperature difference— Hot and Cold fluid inlet/℃		
	Day 4	Day 5	Day 6	Day 4	Day 5	Day 6
10:00	32.495	—	35.61	2.05	—	8.55
11:00	37.468	38.74	38.23	4.25	10.56	10.50

Continued Table 2.9

Hour of day	Hot water inlet temperature/℃			Temperature difference—Hot and Cold fluid inlet/℃		
	Day 4	Day 5	Day 6	Day 4	Day 5	Day 6
12:00	40.119	40.45	38.79	5.85	10.16	10.51
13:00	41.300	38.32	39.09	6.33	9.61	10.07
14:00	40.579	34.11	38.75	6.32	6.96	9.11
15:00	37.868	32.59	36.70	5.86	5.35	8.05
16:00	34.559	29.93	33.06	4.21	4.49	6.27
17:00	27.692	27.63	29.71	2.31	3.31	4.75
Average	36.51	34.54	36.24	4.65	7.20	8.48

The measured cold supply water flow rates from Day 4 to Day 6 in the experiment are as listed in Table 2.10. The supply water was stopped at 10a.m on Day 5, and the experiment results at 10am would be excluded from the validation test.

Table 2.10 Measured cold water flow rate from Day 4 to Day 6 (10:00~17:00)

Time		10	11	12	13	14	15	16	17
Flow rate /mL·min^{-1}	Day 4	190	160	140	140	140	170	180	170
	Day 5	—	200	160	120	180	180	180	180
	Day 6	210	180	180	180	165	140	180	190

The temperature increase of the water layer was still hysteretic compared to that of the outer surface temperature as shown in Fig. 2.12. This phenomenon was the same as that of the water flow window.

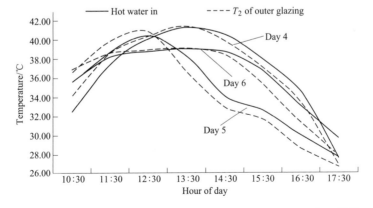

Fig. 2.12 Variation of hot water inlet and T_2 of outer glazing from Day 4 to Day 6 (10am to 6pm)

However, the temperature increase at the center point of outer glazing was not as fast as those of the water flow window on Day 5 and Day 6. The differences between T_2 and

T_3 were larger than those of the water flow window, though the solar radiation was stronger. This was because that the outdoor environment was cooler during these two days, which was about 4℃ lower than that of Day 3 under the similar solar condition. Thus, part of the thermal energy was released back to the ambient. The recorded hourly-averaged temperature of the outer glazing is shown in Fig. 2. 13. The condition for Day 4 was quite different; in that the temperature at the center point was almost equal to the top surface temperature. This was caused by the extremely strong solar radiation at the average value of 468.78W/m². As for the inner glazing temperature on that day, the enhanced convective heat transfer further minimized the difference between T_1 and T_3, and thus the temperature at the center point was quite close to the bottom surface temperature shown in Fig. 2. 14.

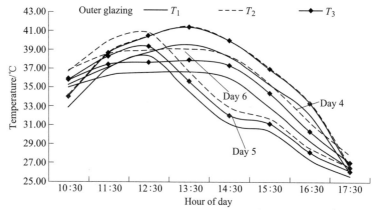

Fig. 2. 13 Surface temperature of outer glazing at different heights from Day 4 to Day 6 (10am to 6pm)

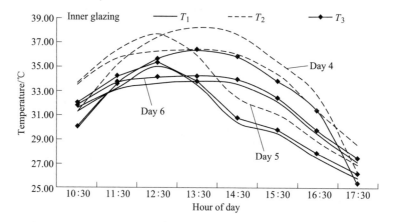

Fig. 2. 14 Surface temperature of inner glazing at different heights from Day 4 to Day 6 (10am to 6pm)

The inlet and outlet temperatures of hot and cold fluids in the heat exchanger are shown in Fig. 2.15 and Fig. 2.16 respectively. Similarly, a stronger solar radiation was followed by a larger temperature increase of cold water and a larger difference between the outlet temperatures of the two fluids stream. The largest difference between the outlet temperatures of the two fluids on Day 4 was 1.8℃, but this difference could be as large as about 4℃ on Day 5 and Day 6. Thermal energy carried by the hot fluid was not transferred to the cold water effectively. The possible cause could be the influence of the anti-freezing fluid, which resulted in lower heat transfer coefficient. As mentioned before, the daily average temperature difference between the hot fluid and cold water inlets on Day 4 was too small and the heat exchange between the two fluids would not be as good as the other cases. The influence of the anti-freezing fluid on water heat gain would be further discussed later.

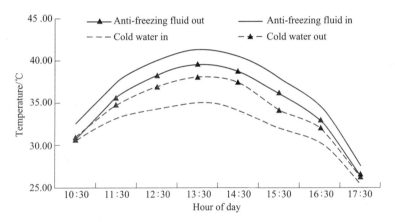

Fig. 2.15 Fluid temperature in the heat exchanger on Day 4 from 10am to 6pm

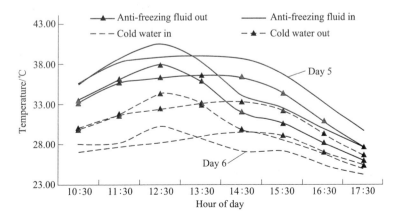

Fig. 2.16 Fluid temperature in the heat exchanger on Day 5 and Day 6 from 10am to 6pm

The summary of daily water heat gain and thermal efficiency for the experiment of anti-freezing liquid filled window are listed in Table 2.11. The thermal efficiency remained high when water was replaced by the anti-freezing fluid. This could be caused by the strong solar radiation and a lower cold water inlet temperature during the experiment periods. The overall thermal efficiency level of the experiment was lower than that of the water flow window experiment, though the solar radiation level was much higher. This was influenced by the higher viscosity and lower thermal capacity. Another cause was expected the more heat release to the ambient because of the cold weather. The extremely low efficiency on Day 4 was caused by its extremely high cold water inlet temperature in the heat exchanger as mentioned before.

Table 2.11 Daily water heat gain and thermal efficiency of anti-freezing liquid flow window of the 3 test days

Date	28th Sep.	2nd Oct.	3rd Oct.
Water heat gain/J	150.82	205.56	298.63
Thermal efficiency/%	4.65	7.48	9.46

2.2.3 Performance Comparison of Water and Anti-freezing Liquid Flow Window

In this experimental study, it was difficult to investigate the effect of liquid properties systematically, for the reason that no comparison experiment was carried out at the same time with the two different working fluids. However, the window systems were under similar working conditions on Day 3 (with water as working fluid) and Day 4 (with anti-freezing fluid as working fluid). Through these, integrative effects of solar radiation, air temperature and wind can be readily assessed. The data acquired on these two days were used for comparison. The solar radiation at 1pm on Day 3 and Day 4 were 534.4W/m^2 and 532.2W/m^2 respectively. And the corresponding thermal efficiencies were 6.95% and 6.6% with a difference of about 5%. The hourly average solar radiation on Day 3 and Day 4 were 335.02W/m^2 and 377.76W/m^2 with daily average thermal efficiency of 8.3% and 5.53%. It could be inferred that the use of anti-freezing liquid might affect the system performance negatively.

However, how the replacement of water with anti-freezing liquid affects the system performance needs to be further discussed in the future by carrying out experiments with different working fluids under the same condition or with numerical method. The results can then be more accurate and reliable.

2.3 Summary

Thermal performance of liquid flow window was evaluated experimentally in this Chapter. And different working fluids of water and anti-freezing liquid were filled into the closed circulation chamber. This is for the potential application of this technique in different climate conditions. The major conclusions are drawn below:

(1) The visual effect of anti-freezing liquid flow window is still good as compared to that of water flow window; this can be observed from Fig. 2.5 and Fig. 2.11. The lux level in the test room should be measured for accurate evaluation of the effect of anti-freezing liquid on visible transmittance.

(2) There exists time delay for heat transfer from outer glazing to liquid layer and inner glazing, and this is observable for both water flow and anti-freezing liquid flow window. A strong solar radiation causes a faster increase in the outer glazing temperature and a slower increase in the inner glazing temperature, the temperature increase of outer glazing is also affected by the outdoor air temperature.

(3) The naked supply water pipe causes a higher inlet temperature of cold feed water, and consequently a decrease of temperature difference between hot fluid and cold feed water at inlets. This small temperature difference affects the heat transfer coefficient of the heat exchanger negatively. From the point of system thermal efficiency improvement, using of good thermal insulation on the supply water pipe is necessary. However, the supply water can be heated directly by exposing the water pipe to the solar radiation, especially in the summer season. And this is also good from the point of energy harvesting and utilization.

(4) The heat transfer coefficient in the heat exchanger of anti-freezing liquid flow window may not be as good as that of water flow window. But the way how the anti-freezing liquid influences the heat transfer coefficient and the overall system performance should be further studied by experimental or numerical methods.

3 Schematic Optimization of Liquid Flow Window

How to make better use of solar energy in order to reduce carbon emission and its global impact on climate change has been receiving worldwide attention. Since its introduction, the novel idea of 'multi-glazed water flow window' is now under the investigation of several research teams. Thermal performance of this novel technique was found affected by weather conditions, liquid properties and its flow velocity, as well as the integration with alternative energy systems. And few of the studies found that the impacts of glazing properties were also significant. Under the forced circulation scheme, the effect of water flow rate was found particularly evident at the low velocity range. For the buoyant-driven system, the flow rate is affected much by the pipe-work design including the distribution headers setting and the pipe size, as a result of the change in friction loss.

The importance of proper window structure design for good energy performance has been pointed out in related research articles. The solar heat gain coefficient (g) of glazing system is related closely to the water flow rate. While a stagnant water layer will result in maximum g value, adequate water flow is able to reduce the energy transfer both inwards and outwards. Besides, geographical factors like the sun path and solar intensity are amongst the key influencing factors on thermal performance, so are the major dimensions of the window components.

From the above, one can observe that the research works in the past were more or less focusing on studying the impact of external influencing factors or major operating parameters on the window system performance. Very few were evaluating the window component design and configuration in details. In fact, there can be further improvement of this new technology to adapt to various applications. For example, the presence of the water-filled tubing and liquid volume increases the overall weight of the window. This adds difficulties to the construction, site installation and future replacement tasks, as well as increases the imposed loads on the building structure and the overall investment costs. Thus, in this chapter, the schematic optimization is introduced. The effects of the height to width ratio with the same glazing area and the thickness of cavity space, as well as the pipe-work design were evaluated.

Heat transfer and fluid flow pattern at the glazed façade have been widely studied in both experimental and numerical approaches. Claros-Marfil et al. studied water-flow window with an open source data acquisition and controller; reliable test results were obtained. Accurate results were also obtained via numerical analysis, like in the study of Ismail et al. with both one- and two-dimensional heat transfer models. Xamán et al., developed a computer program in FORTRAN to predict the thermal performance of double-glazing window; good agreement with experimental measurements was demonstrated. The schematic optimization study in this chapter will be carried out numerically with the same FORTRAN simulation approach.

3.1 Research Methodology

In the current study, one-dimensional water flow in the window cavity was assumed in developing the thermal simulation model, since laminar flow was expected under the buoyance driven force. Numerical models for fluid flow and energy balance of the system components were developed and discretized with the finite difference explicit approach. The simulation study was completed with a self-developed FORTRAN program. With these, year-round dynamic system simulations on hourly basis were carried out for energy performance evaluation.

In the simulation, indoor air temperature was assumed to be 21℃ from Nov. to Apr. (the next year) and 25℃ in the remaining months; the cold feed water was at a steady flow rate of 200mL/min and the inlet temperature was taken the same as the ambient temperature. The annual performance evaluation was completed with the use of the TMY weather data of China. The system performance was compared from several aspects: room heat gain and heat loss through the window glazing, cold water heat gain and thermal efficiency, as well as electricity saving.

3.1.1 Mathematical Models Development

Energy flow paths at the window surface are given in Fig. 3.1. The incident solar radiation is partly reflected, partly transmitted and partly absorbed by the window components and the water layer. Convection and radiation heat transfer exist between the outermost/innermost glazing surfaces and the outdoor/indoor environment. Heat transfer in the glazing panes is conduction dominated, and heat exchange between the glazing surface and the water layer is realized by thermal convection.

Based on the above, numerical models for flow and heat transfer calculation were

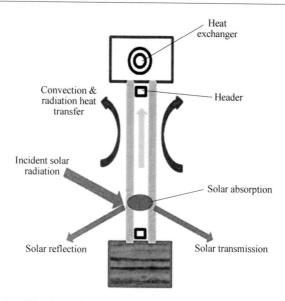

Fig. 3.1 Side view of water flow window indicating the energy flow paths

developed considering several aspects, i. e., determination of solar radiation on the tilted surface, liquid flow velocity, heat transfer rates in the liquid layer, at the two glazing panes, and in the heat exchanger. The developed mathematical models are given below.

3.1.1.1 Solar Radiation

The solar radiation incident on the tilted surface could be calculated based on the global radiation incident on a horizontal surface. With isotropic sky, the total solar radiation on a tilted surface is calculated by

$$G_{tt} = G_{bh}R_b + G_{dh}\left(\frac{1 + \cos\beta}{2}\right) + G_{th}\gamma_g\left(\frac{1 - \cos\beta}{2}\right) \quad (3\text{-}1)$$

Where G_{tt} ——the total solar radiation incident on a tilted surface in W/m²;

G_{th} ——the total solar radiation incident on a horizontal surface in W/m²;

G_{bh} ——the beam radiation incident on a horizontal surface in W/m²;

G_{dh} ——the diffuse radiation incident on a horizontal surface in W/m²;

γ_g ——the reflectance of ground;

β ——the slope of the surface, $0 \leqslant \beta \leqslant 180°$;

R_b ——the ratio of beam radiation on the tilted surface to that on a horizontal surface, given by

$$R_b = \frac{\cos\theta}{\cos\theta_z} \quad (3\text{-}2)$$

Where θ ——the angle of incidence in(°);

θ_z ——the zenith angle, which means the angle between the vertical and the line to the sun; it equals to the angle of incidence of beam radiation for a horizontal surface in(°).

The values of θ and θ_z are calculated by

$$\cos\theta = \sin\delta\sin\phi\cos\beta - \sin\delta\cos\phi\sin\beta\cos\gamma + \cos\delta\cos\phi\cos\beta\cos\omega +$$
$$\cos\delta\sin\phi\sin\beta\cos\gamma\cos\omega + \cos\delta\sin\beta\sin\gamma\sin\omega \quad (3\text{-}3)$$

$$\cos\theta_z = \cos\delta\cos\phi\cos\omega + \sin\delta\sin\phi \quad (3\text{-}4)$$

Where, δ is the declination, which is the angular position of the sun at solar noon, in that $-23.45 \leqslant \delta \leqslant 23.45$ with north positive, and it is given by

$$\delta = 23.45\sin\left(360\,\frac{284+n}{365}\right) \quad (3\text{-}5)$$

Where n ——the day of year;

ϕ ——the angular latitude with north positive, $-90° \leqslant \phi \leqslant 90°$;

γ ——the surface azimuth angle, with zero due south, east negative, and west positive, $-180° \leqslant \gamma \leqslant 180°$;

ω ——hour angle, with morning negative and afternoon positive, given by

$$\omega = 15(H - 12) \quad (3\text{-}6)$$

H ——the time of a day in hour from 1 to 24.

The total radiation on a horizontal surface can be split into beam and diffuse components. The commonly used approach is to calculate the fraction of the hourly diffuse radiation to the global radiation on a horizontal surface in relation to the hourly clearness index. The hourly diffuse fraction on horizontal surface is calculated by

$$\frac{G_{dh}}{G_{th}} = \begin{cases} 1.0 - 0.249 k_T, & k_t < 0.35 \\ 1.557 - 1.84 k_T, & 0.35 < k_t < 0.75 \\ 0.177, & 0.75 < k_t \end{cases} \quad (3\text{-}7)$$

The hourly clearness index k_T can be calculated by

$$k_T = \frac{G_{th}}{G_o} \quad (3\text{-}8)$$

Where G_o ——the extraterrestrial radiation on a horizontal surface, given by

$$G_o = G_{sc}\left(1 + 0.33\cos\frac{360n}{365}\right)\cos\theta_z \quad (3\text{-}9)$$

Where G_{sc} ——the solar constant with value of 1367 W/m².

The diffuse radiation on the tilted surface can also be calculated with a detailed model considering the isotropic diffuse, the circumsolar diffuse and the diffuse from the horizon. The sky is considered to be an anisotropic sky under this circumstance. The diffuse radiation on a tilted surface is calculated by

$$G_{dt} = G_{dh}\left[(1-F_1)\left(\frac{1+\cos\beta}{2}\right) + F_1\frac{a}{b} + F_2\sin\beta\right] \quad (3\text{-}10)$$

Where G_{dt} ——the diffuse radiation incident on a tilted surface with tilt angle of β in W/m^2;

F_1, F_2 ——the circumsolar and horizon brightness coefficients respectively, given by

$$F_1 = \max\left[0, \left(f_{11} + f_{12}\Delta + \frac{\pi\theta_z}{180}f_{13}\right)\right] \quad (3\text{-}11)$$

and

$$F_2 = f_{21} + f_{22}\Delta + \frac{\pi\theta_z}{180}f_{23} \quad (3\text{-}12)$$

Where Δ —— the sky brightness, calculated by

$$\Delta = m\frac{G_{dh}}{G_{on}} \quad (3\text{-}13)$$

Where m ——the air mass calculated approximately by

$$m = 1/\cos\theta_z \quad (3\text{-}14)$$

This is for zenith angle from 0° to 70°. More discussions on the value of m with higher zenith angle can be found in. G_{on} is the extraterrestrial normal incidence radiation given by

$$G_{on} = G_{sc}\left(1 + 0.33\cos\frac{360n}{365}\right) \quad (3\text{-}15)$$

The values of f_{11}, f_{12}, f_{13}, f_{21}, f_{22} and f_{23} can be determined according to the ranges of sky clearness ε, which is calculated by

$$\varepsilon = \frac{\dfrac{G_{dh} + G_n}{G_{dh}} + 5.535\times10^{-6}\theta_z^3}{1 + 5.535\times10^{-6}\theta_z^3} \quad (3\text{-}16)$$

where G_n ——the normal incidence beam radiation calculated by

$$G_n = \frac{G_{bh}}{\cos\theta_z} \quad (3\text{-}17)$$

Also in equation (3-10), a and b are the terms that account for the angles of incidence of the cone of circumsolar radiation on the tilted and horizontal surfaces. They are given by

$$a = \max[0, \cos\theta] \quad (3\text{-}18)$$

and

$$b = \max[\cos 85, \cos\theta_z] \quad (3\text{-}19)$$

where, the ratio of a/b becomes R_b for most hours when the collectors will have useful outputs.

The solar radiation incident on the window surface is usually divided into three parts as

$$G_{tt} = G_r + G_t + G_a \qquad (3\text{-}20)$$

Where G_r ——the reflected solar radiation in W/m^2;

G_t ——the transmitted solar radiation in W/m^2;

G_a ——the absorbed solar radiation by each component of the system, including the two glazing panes and the water layer in W/m^2.

To study the heat transfer mechanism at the water layer and the two glazing panes, 2D energy balance equations of each component were built, as described in the following sections.

3.1.1.2 Outer glazing

The heat source of the outer glazing comes mainly from the following: the absorption of solar thermal energy, the convection and radiation heat transfer with the ambient, and the convection heat transfer with the liquid layer. Mathematical model for the thermal balance is thus given as

$$\rho_{out} C_{out} D_{out} \frac{\partial T_{out}}{\partial t} = \alpha_{out} G + \left(\frac{1}{1/h_{c,1} + 0.5 D_{out}/k_{out}} \right) (T_a - T_{out}) + h_{r,1}(T_a - T_{out}) +$$

$$\left(\frac{1}{1/h_{f,2} + 0.5 D_{out}/k_{out}} \right) (T_f - T_{out}) + k_{out} D_{out} \frac{\partial^2 T}{\partial y^2} \qquad (3\text{-}21)$$

Where ρ_{out} ——the density of the outer glazing in kg/m^3;

C_{out} ——the specific heat capacity of the outer glazing in $J/(kg \cdot K)$;

D_{out} ——the thickness of the outer glazing in m;

k_{out} ——the thermal conductivity of the outer glazing in $W/(m \cdot K)$;

α_{out} ——the thermal absorption coefficient of the outer glazing;

G——the solar radiation incident on unit area of glazing surface in W/m^2;

$h_{c,1}$ ——the convection heat transfer coefficient between the outer glazing and the ambient in $W/(m^2 \cdot K)$;

$h_{r,1}$ ——the radiation heat transfer coefficient between the outer glazing and the ambient in $W/(m^2 \cdot K)$;

$h_{f,2}$ ——the convection heat transfer coefficient between the outer glazing and the liquid layer in $W/(m^2 \cdot K)$;

T_a, T_{out}, T_f ——the temperatures of ambient, outer glazing and liquid layer, respectively in ℃.

3.1.1.3 Inner Glazing

Similar to the heat transfer mechanism of the outer glazing, the heat transfer at the inner glazing is also calculated based on the absorption of solar thermal energy, the convection and radiation heat transfer with the indoor air, and the convection heat transfer with the liquid layer. Hence,

$$\rho_{in} C_{in} D_{in} \frac{\partial T_{in}}{\partial t} = \alpha_{in} G + \left(\frac{1}{1/h_{c,4} + 0.5 D_{in}/k_{in}}\right)(T_r - T_{in}) + h_{r,4}(T_r - T_{in}) + \left(\frac{1}{1/h_{f,3} + 0.5 D_{in}/k_{in}}\right)(T_f - T_{in}) + k_{in} D_{in} \frac{\partial^2 T}{\partial y^2} \quad (3\text{-}22)$$

Where ρ_{in} ——the density of the inner glazing in kg/m³;
  C_{in} ——the specific heat capacity of the inner glazing in J/(kg·K);
  D_{in} ——the thickness of the inner glazing in m;
  k_{in} ——the thermal conductivity of the inner glazing in W/(m·K);
  α_{in} ——the thermal absorptivity of the inner glazing;
  $h_{c,4}$ ——the convection heat transfer coefficient between the inner glazing and the indoor environment in W/(m²·K);
  $h_{r,4}$ ——the radiation heat transfer coefficient between the inner glazing and the indoor environment in W/(m²·K);
  $h_{f,3}$ ——the convection heat transfer coefficient between the inner glazing and the liquid layer in W/(m²·K);
  T_r, T_{in} ——the temperatures of indoor air and inner glazing respectively in ℃.

3.1.1.4 Water Layer

For the water layer, the heat source comes from the direct thermal absorption of solar radiation and the convection heat transfer with the outer and inner glazing panes. The conduction heat transfer along the vertical direction is neglected. The mathematical model for the thermal balance of the water layer is written as

$$\rho_f C_f D_f \left(\frac{\partial T_f}{\partial t} + u_f \frac{\partial T_f}{\partial y}\right) = \alpha_f G + \left(\frac{1}{1/h_{f,2} + 0.5 D_{out}/k_{out}}\right)(T_{out} - T_f) + \left(\frac{1}{1/h_{f,3} + 0.5 D_{in}/k_{in}}\right)(T_{in} - T_f) \quad (3\text{-}23)$$

Where ρ_f ——the density of liquid layer in kg/m³;
        C_f ——the specific heat capacity of liquid layer in J/(kg · K);
        D_f ——the thickness of liquid layer in m;
        α_f ——the thermal absorptivity of liquid layer.

3.1.1.5 Heat Transfer Coefficient

Heat transfer coefficients at the glazing surfaces are calculated by

$$h_{c,1} = 2.8 + 3.0v \tag{3-24}$$

$$h_{r,1} = \frac{\sigma(\Theta_a^2 + \Theta_1^2)(\Theta_a + \Theta_1)}{\dfrac{1}{\varepsilon_a} + \dfrac{1}{\varepsilon_1} - 1} \tag{3-25}$$

$$h_{c,4} = 4.3 \tag{3-26}$$

$$h_{r,4} = \frac{\sigma(\Theta_4^2 + \Theta_r^2)(\Theta_4 + \Theta_r)}{\dfrac{1}{\varepsilon_4} + \dfrac{1-\varepsilon_r}{\varepsilon_r(2WH + 2WL + LH)}} \tag{3-27}$$

Where v ——the wind velocity in m/s;
$\Theta_a, \Theta_1, \Theta_4, \Theta_r$ ——the temperatures of outdoor air, surface of outer glazing facing outdoor, surface of inner glazing facing the indoor environment and the indoor air respectively in ℃;
$\varepsilon_a, \varepsilon_1, \varepsilon_4, \varepsilon_r$ ——the emissivity of outdoor air, surface of outer glazing facing the outdoor, surface of inner glazing facing the indoor environment, and the room surface respectively;
W, L, H ——the width, length and height of the test chamber respectively in m.

Heat transfer between the flowing fluid and the two glazing surfaces is calculated by

$$h_{f,2} = \frac{N_u \cdot k}{D} \tag{3-28}$$

Where k ——the conductivity of working fluid in W/(m · K);
        D ——the hydraulic diameter in m;
        N_u ——the Nusselt number which is calculated differently for the entrance and the fully-developed part, calculated by

$$N_u = 1.86\,(RePr)^{\frac{1}{3}} \left(\frac{D}{L}\right)^{\frac{1}{3}} \left(\frac{\mu}{\mu_w}\right)^{0.14} \tag{3-29}$$

Where Re, Pr ——the Reynolds and Prandtl numbers of water layer;
        L ——the length of flow path in m;

μ, μ_w ——the kinetic viscosity of glazing and water respectively.

For the fully developed part, we obtain

$$N_u = 7.54 \qquad (3\text{-}30)$$

3.1.1.6 Liquid Flow Velocity in the Closed Circuit

The flow of the liquid layer is driven by the thermo-syphon effect. The driven force P_T should be equal to the friction loss P_f, written as

$$P_T = P_f \qquad (3\text{-}31)$$

and

$$P_T = (\rho_{down} h_{down} - \sum \rho_{up} h_{up}) g \qquad (3\text{-}32)$$

Where ρ_{down} ——the density of down flow liquid from the outlet of heat exchanger to the inlet of window cavity in kg/m³;

h_{down} ——the corresponding height from the outlet of heat exchanger to the inlet of window cavity in m;

ρ_{up} ——the density of up flow water, including the liquid flow from the inlet to the outlet of the window cavity and the liquid flow from the outlet of the window cavity to the inlet of the heat exchanger in kg/m³;

h_{up} ——the corresponding height of the window and the height from the window cavity outlet to the heat exchanger inlet in m;

g ——the gravity constant in kg · m/s.

$$P_f = \sum f \frac{l}{D} \cdot \frac{\rho u^2}{2} + \sum \zeta \frac{\rho u^2}{2} \qquad (3\text{-}33)$$

Where f ——the linear resistance coefficient, given by

$$f = \frac{64}{R_e} = \frac{64}{u \cdot D} \qquad (3\text{-}34)$$

l ——the length of the flow path in m;

D ——the equivalent diameter of the flow path in m;

ρ ——the density of liquid in the corresponding flow path in kg/m³;

u ——the liquid flow velocity in the corresponding flow path in m/s;

ζ ——the local resistance coefficient.

3.1.1.7 Heat Exchanger

The heat transfer at the double-pipe heat exchanger is calculated with the NTU-ε method. Where the hot fluid and cold water is arranged as counter flow, the maximum heat transfer can first be calculated by

$$Q_{max} = (Mc)_{min}(T_{hin} - T_{cout}) \tag{3-35}$$

Where $(Mc)_{min}$ ——the minimum value of the product of mass flow rate and heat capacity of the heat transfer fluid in J/(K·s);

         T_{hin} ——the hot fluid inlet temperature in ℃;

         T_{cout} ——the cold water inlet temperature in ℃.

The number of heat transfer units (NTU) is calculated by

$$\text{NTU} = \frac{kA}{C_{min}} \tag{3-36}$$

Where A ——the heat transfer area in m²;

         k ——the heat transfer coefficient of the heat exchanger in W/(m²·K).

$$C_{min} = (Mc)_{min} \tag{3-37}$$

$$C_{max} = (Mc)_{max} \tag{3-38}$$

$$C = \frac{C_{min}}{C_{max}} \tag{3-39}$$

The effectiveness of the heat exchanger is calculated by

$$\varepsilon = \begin{cases} \dfrac{3 - e^{-\text{NTU}(1-C)}}{1 - C \cdot e^{-\text{NTU}(1-C)}}, & C \neq 1 \\ \dfrac{\text{NTU}}{1 + \text{NTU}}, & C = 1 \end{cases} \tag{3-40}$$

The actual amount of energy exchanger between two fluids is calculated by

$$Q = Q_{max}\varepsilon \tag{3-41}$$

The outlet temperatures of hot fluid and cold water can then be calculated by

$$Q = Mc\Delta T_{out-in} \tag{3-42}$$

Where ΔT_{out-in} ——the temperature difference of hot fluid or cold water between the inlet and the outlet of the heat exchanger.

3.1.2 Model Validation

3.1.2.1 Uncertainty Analysis

The agreement between experimental and simulation results is affected by the accuracy of measuring instrument in the experiment and the numerical method in use. This makes it necessary to carry out uncertainty analysis to find out their possible effects on the outputs. The process is to find out the possible difference between the experimental and simulation results caused by the error in measurement and the uncertainty in inputs. The model validation is considered to be successful if the deviations between the experiment and simulation findings are within the uncertainty range. There are many approaches that can be adopted in uncertainty test, including the internal and external methods. The factorial method adopted in this study is one of the external methods, which takes the simulation as a 'black box' and changes only the inputs. Besides, the interaction of different influence factors can be well modeled with this method.

In uncertainty analysis, factors with large uncertainty and those that have significant impact on the outputs should be decided in the first place. All the possible values of the selected influential factors and their interaction are taken into consideration. Two variables of X and Y with two and three possible values are used to illustrate this as shown in Fig. 3. 2. In the case with two possible values for both X and Y, there can be $2^2 = 4$ possible outputs and the corresponding number of outputs of the case with three possible values for both X and Y should be $3^2 = 9$ taking into consideration their interaction. And the total number of possible outputs can be defined as $Z = P^N$, in which P means the possibilities of each factor and N represent the number of factors considered. This is true when all the factors have the same number of possible values. It can be generalized as $Z = P_1 \times P_2 \times \cdots \times P_{n-1} \times P_n$ with 1 to n means the first to the last variables in sequence and P is the number of the possible values of the corresponding variables.

	X+	X−
Y+	Z(X+,Y+)	Z(X−,Y+)
Y−	Z(X+,Y−)	Z(X−,Y−)

(a)

	X+	X	X−
Y+	Z(X+,Y+)	Z(X,Y+)	Z(X−,Y+)
Y	Z(X+,Y)	Z(X,Y)	Z(X−,Y)
Y−	Z(X+,Y−)	Z(X,Y−)	Z(X−,Y−)

(b)

Fig. 3. 2 Number of outputs with two factors
(a) Two possible values for each factor(2^2); (b) Three possible values for each factor(3^2)

The accuracy of the outputs in this study is mainly affected by the errors in measurement and the empirical-based calculations of heat transfer coefficients. Seven factors were considered and three values were used for each of them; a standard one from measurement or formula, and a larger and a smaller one. The uncertainty ranges were from the instrument supplier and a PhD thesis. Thus, there should be totally $3^7 = 2187$ cases taking into consideration the variation and interaction of the seven factors, and the error bands can be decided based on the outputs of the 2187 cases. Those factors considered and their corresponding uncertainty ranges are given as:

(1) Measured solar radiation on the horizontal surface, with uncertainty of 4%;
(2) Air temperature, with uncertainty of 0. 5 ℃;
(3) Cold water inlet temperature, with uncertainty of 0. 5 ℃;
(4) Cold water velocity, with uncertainty of 4%;
(5) Wind velocity, with uncertainty of 4%;

(6) Convection heat transfer coefficient between the outer glazing surface and the outdoor environment, with uncertainty of 4%;

(7) Convection heat transfer coefficient between inner glazing surface and indoor air, with uncertainty of 4%.

Sensitivity analysis was conducted based on the testing conditions at noon. The variation of hot water inlet and cold water outlet temperatures in the heat exchanger, and glazing surface temperatures of the 2187 cases at noon on Day 1 in the experiment of water flow window were calculated. They are given in Fig. 3.3.

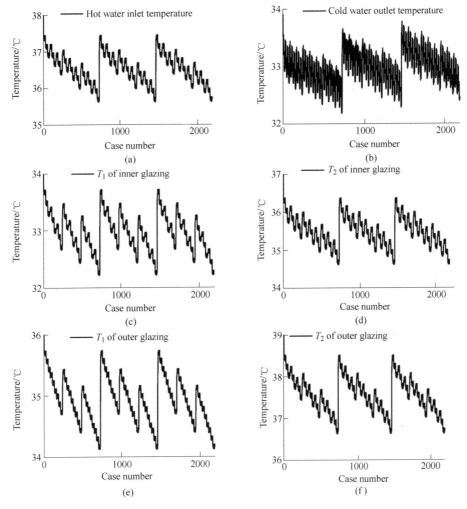

Fig. 3.3 Variation of temperature at noon on 25th September for the 2187 cases
(a) Hot fluid inlet temperature; (b) Cold water outlet temperature; (c) T_1 of inner glazing;
(d) T_2 of inner glazing; (e) T_1 of outer glazing; (f) T_2 of outer glazing

The measured values of T_1 and T_2 of the inner and outer glazing were 33.1℃, 35.5℃, 34.3℃ and 36.7℃, respectively; and the corresponding temperature ranges in the uncertainty analysis were 32.2~34.3℃, 34.6~37.1℃, 34.1~36.4℃ and 36.6~39.3℃, respectively. For the temperatures of the hot water inlet and cold water outlet, the measured values were 37.6℃ and 33.9℃ respectively. The corresponding temperature ranges in the sensitivity analysis were from 35.6℃ to 38.3℃ and from 32.2℃ to 33.91℃, respectively.

All the temperatures measured in the experiment located well within the range calculated by simulation. And the error bands were ±1.83℃ for hot water inlet temperature, ±1.36℃ for cold water outlet temperature in the heat exchanger; ±1.57℃ and ±1.76℃ for T_1 and T_2 of the inner glazing; and ±1.62℃ and ±1.82℃ for T_1 and T_2 of the outer glazing. It was calculated by one half of the temperature range in simulation plus the measurement error of the thermocouple.

3.1.2.2 Validation Results

The validation exercise was conducted based on the daily experiment results from 10 am to 6pm. The calculated water and glazing temperatures from the self-developed FORTRAN program were compared with the experimental test data with water as working fluid. They are shown in Fig. 3.4 to Fig. 3.6 in which the simulation and experimental results are abbreviated as 'Sim' and 'Exp' for short. The comparison shows good agreement between the two sets of results, with all the temperatures located within the allowable error bands.

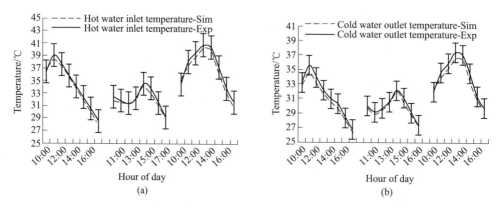

Fig. 3.4 Comparison of water temperatures in experiment and FORTRAN computation from 10 am to 6pm for the water flow window

(a) Hot water inlet; (b) Cold water outlet

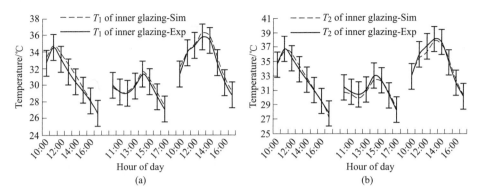

Fig. 3.5 Comparison of inner glazing surface temperatures in experiment and FORTRAN computation from 10 am to 6pm for the water flow window
(a) T_1; (b) T_2

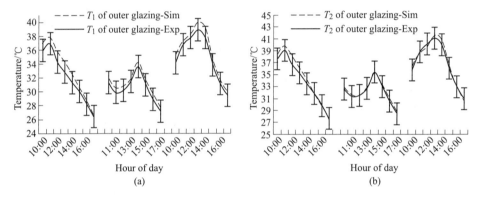

Fig. 3.6 Comparison of outer glazing surface temperatures in experiment and FORTRAN computation from 10 am to 6pm for the water flow window
(a) T_1; (b) T_2

3.2 Performance Evaluation with the Variation of Schematic Parameters

3.2.1 Thermal Performance at Different Cavity Spaces

In the simulation study, the window dimension is 1.2m(H)×0.8m(W), and the heat exchanger is of the same dimension as used in the experiment with outer and inner pipe diameters of 30mm and 20mm. Three more cases including 20mm, 15mm and 10mm water layer thickness were considered besides the previously used 30mm water layer. Also, absorptive glazing with thermal and optical properties listed in Table 3.1 was employed.

Table 3.1 Variation of optical properties of absorptive glazing with incident angle

Incident angle/(°)	0	10	20	30	40	50	60	70	80
Transmittance	0.307	0.305	0.299	0.290	0.276	0.257	0.229	0.184	0.102
Absorptance	0.645	0.645	0.651	0.659	0.669	0.675	0.670	0.630	0.493
Visible transmittance	0.461	0.459	0.453	0.443	0.426	0.402	0.364	0.297	0.166
Reflectance	0.048	0.050	0.050	0.051	0.055	0.068	0.101	0.186	0.405

Thermal absorption coefficient of the 10mm water layer is 0.1357; this is close to the absorption coefficient of clear glass with a thickness of 2mm. And the equivalent absorption coefficients of water layer with different thickness were calculated with the WINDOW software accordingly. The thermal absorption coefficients with a range of water layer thickness are given in Table 3.2.

Table 3.2 Thermal absorption coefficients of water layer with different thickness

Water thickness/mm	10	15	20	30
Thermal absorption	0.1357	0.162	0.187	0.228

The temperature increase was found changed with the water layer thickness, as the thermal resistance was changed accordingly. The hot water inlet temperature in the heat exchanger which is taken the same as the outlet temperature of the window cavity on 1st Jan. from 10 am to 6pm is presented in Fig. 3.7 as an illustration. With a thinner water layer, the water temperature in the top window zone increased faster in the morning and the peak temperature was also higher. The difference between the peak temperatures of the two cases with 10mm and 30mm water layer was 2.58℃. However, the condition was quite different in the afternoon. The water temperature of the case with larger water layer

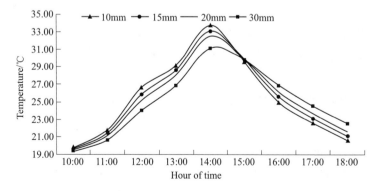

Fig. 3.7 Hot water inlet temperature in the heat exchanger on 1st Jan from 10 am to 6pm

thickness was found higher because of the thermal energy storage in the water layer. And the water temperature was more stable throughout the day with larger amount of water filled in the cavity. The water temperature affected both the heat transfer rates in the heat exchanger and at the two glazing panes.

The water heat gain is influenced by several factors, including water temperature increase, water flow rate and heat exchanger effectiveness. The monthly water heat gains brought about by the 0.96m² glazing are listed in Table 3.3 and its variation is shown in Fig. 3.8.

Table 3.3 Monthly water heat gain at different water layer thicknesses

(kW·h)

Month	Case 1(10mm)	Case 2(15mm)	Case 3(20mm)	Case 4(30mm)
Jan.	11.01	10.89	10.82	10.46
Feb.	8.46	8.40	8.30	8.05
Mar.	7.11	7.04	6.94	6.69
Apr.	5.33	5.24	5.13	4.86
May	6.29	6.18	6.06	5.76
Jun.	4.88	4.78	4.67	4.41
Jul.	5.95	5.83	5.70	5.37
Aug.	6.56	6.44	6.31	5.98
Sep.	6.22	6.13	6.02	5.77
Oct.	9.89	9.79	9.67	9.36
Nov.	10.99	10.90	10.80	10.54
Dec.	12.57	12.44	12.15	11.75
Sum	95.27	94.06	92.56	89.00

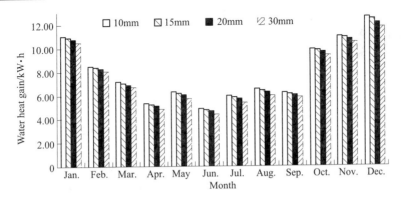

Fig. 3.8 Variation of water heat gain with different water layer thickness

The yearly total water heat gains were 95.27kW·h, 94.06kW·h, 92.56kW·h and 89kW·h for the cases with water layer of 10mm, 15mm, 20mm and 30mm, respectively. The increase in water heat gain was more evident when the water layer thickness was reduced from 30mm to 20mm. This is at 3.56kW · h for the 0.96m² glazing surface, corresponding to an increase of 4.09%. The increase in annual water heat gain was less than 2% when the water layer thickness was further reduced to 15mm and 10mm. In the winter season, the solar radiation was strong, and the system performance was more sensitive to the variation in water layer thickness.

The monthly and yearly averaged system thermal efficiencies are listed in Table 3.4 and the variation of thermal efficiency is shown in Fig. 3.9. The yearly average thermal efficiency increased with the decrease of water layer thickness as expected. Also, the increase of average thermal efficiency was larger with the water layer thickness decreased

Table 3.4 System thermal efficiency with variation in water layer thickness

(%)

Month	Case1(10mm)	Case2(15mm)	Case3(20mm)	Case4(30mm)
Jan.	14.54	14.38	14.28	13.80
Feb.	13.57	13.47	13.32	12.92
Mar.	11.69	11.56	11.40	11.00
Apr.	8.74	8.58	8.40	7.96
May	10.09	9.92	9.73	9.25
Jun.	8.05	7.89	7.70	7.27
Jul.	7.73	7.57	7.39	6.97
Aug.	8.37	8.22	8.04	7.63
Sep.	8.31	8.18	8.04	7.71
Oct.	10.68	10.58	10.45	10.12
Nov.	12.87	12.77	12.65	12.34
Dec.	13.91	13.77	13.44	13.00
Average	10.71	10.57	10.40	10.00

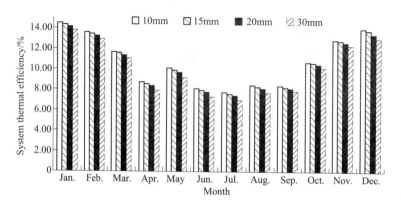

Fig. 3.9 Variation of thermal efficiency with different water layer thickness

3.2 Performance Evaluation with the Variation of Schematic Parameters

from 30mm to 20mm, and the increase in thermal efficiency was decreased with the further reduction in water layer thickness.

Desirable system performance should be evaluated in an integrated manner considering not only the useful water heat gain, but also the effect on air-conditioning load. The glazing surface temperature could be different at various water layer thicknesses, and the room air-conditioning load was related directly to the inner glazing surface temperature. The yearly water heat gains, as well as room heat gain and losses through the window are given in Table 3.5. And the variations of room heat exchanges through the window are given in Fig. 3.10. The negative value indicates that thermal energy is transmitted outward to the outdoor environment through the window system. Both room heat gains and heat losses increase with the reduction of water layer thickness, since the thermal resistance then becomes smaller. The effect on the increase of room heat gains is more evident when the water layer thickness is reduced from 30mm to 20mm, corresponding to an increase of 9.65kW · h.

Table 3.5 Yearly water heat gains, heating and cooling loads with variation in water layer thickness (kW · h)

Water thickness/mm	10	15	20	30
Solar radiation	805.78	805.78	805.78	805.78
Room heat gains	266.02	260.54	255.29	245.64
Room heat losses	58.40	43.70	42.52	40.05
Total water heat gains	95.27	94.06	92.56	89.00

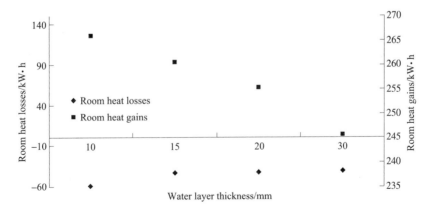

Fig. 3.10 Comparison of annual thermal transmissions through window glazing

The useful water heat gain contributed to energy saving in water heating, while the increase in cooling and heating loads would result in the increase in energy consumption of the air-conditioning system. Thus the overall effect in the variation of water layer thickness should be evaluated based on the overall electricity consumption that includes the electricity consumption of the air-conditioning system less the electricity saving in water heating system. The coefficient of performance (COP) of the air-conditioning system was taken as 3.5 for cooling and 4.5 for heating. The efficiency of electrical water heater was taken as 0.99. The corresponding electricity consumption and saving are listed in Table 3.6. The total energy saving was calculated by deducting the electricity consumption of the air-conditioning system from the electricity saving of water heater. Energy saving potential of the case with 20mm water layer was the largest among all the cases. The electricity saving was 2.7% higher than the case with water layer of 30mm, and the energy saving would be reduced by 2.2% if the water layer thickness was reduced to 15mm. However, the decrease in energy saving could be as large as 33.32% by further reducing the water layer thickness from 15mm to 10mm.

Table 3.6 Electricity consumption and saving at different water layer thickness

(kW · h)

Water thickness/mm	10	15	20	30
Air-conditioning electricity consumption	88.98	84.15	82.39	79.08
Electricity saving of water heater	96.23	95.01	93.49	89.9
Overall electricity saving	7.24	10.86	11.11	10.8

3.2.2 Thermal Performance Comparison with Various Pipe-work Designs

Liquid flow velocity in the window cavity, which is related closely to system performance, is governed by the driving force and the friction loss along the flow path, including local friction loss at the pipe fittings and at the openings of headers, as well as on-way friction loss in the flowing pipe. The impact of pipe work design is evaluated in this section. The window under investigation has a same area of 1.2 m(H)×0.8 m(H), and the double-pipe heat exchanger with diameters of 20mm and 30mm were adopted in the numerical analysis.

Simulations were first conducted to explore the impact of distribution headers, by fixing the cavity space at 20mm as recommended in the previous section. To further evaluate the effects of the connecting-pipe size, the co-variation of cavity space and connecting-pipe diameters were considered. The cavity spaces at 15mm, 20mm and 25mm were selected. The connecting-pipes at three different diameters were also investigated, corresponding to each of the three cavity spaces. Accordingly, the nine test cases listed in Table 3.7 were analyzed.

3.2 Performance Evaluation with the Variation of Schematic Parameters

Table 3.7 Tested cases with different cavity spaces and connecting-pipe diameters

Case number	1	2	3	4	5	6	7	8	9
Cavity/mm	15	15	15	20	20	20	25	25	25
Pipe diameter/mm	10	15	20	15	20	25	20	25	30

3.2.2.1 Impact of the Distribution Headers

The yearly water heat gain and room heat gain of a single piece of water flow window (with or without headers) are given in Table 3.8. By removing the distribution headers, water heat gain is increased by 4.8% because of the enhanced heat transfer. Room heat gain through the window is reduced at the same time, with magnitude less than 1%. Overall speaking, the energy saving potential is promising. A higher yearly average thermal efficiency of 13.58% (as compared to the original 12.96%) is achieved. It is consistent with the experimental findings.

Table 3.8 Thermal performance comparison with different header designs

With/Without header	Water heat gain/kW	Room heat gain/kW	Thermal efficiency/%
With	95.26	206.6	12.96
Without	99.83	205.4	13.58

The rise in water heat gain and the drop in room heat gain are owing to the improved thermal extraction in the cavity, which in turn is caused by the enhanced water circulation. Shown in Fig. 3.11 are the variations in mass flow rate during the typical summer ($2^{nd} \sim 8^{th}$ Sep.) and winter ($8^{th} \sim 14^{th}$ Jan.) weeks for illustration. By eliminating the distribution headers, the mass flow rates are increased for both the summer and winter periods. The increases in water circulation rate are found 36.4% and 57% on average during

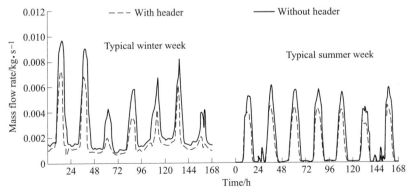

Fig. 3.11 Variations in water circulation rate for window with and without headers

the typical summer and winter weeks, respectively. All these indicate that the water circulation can be enhanced and the energy saving performance of the system can be improved by removing the distribution headers.

3.2.2.2 Impact of the Flow Channel Dimensions

To evaluate the impact of flow channel dimensions on the thermal performance of window without headers, the water heat gains of the nine studied cases with different connecting-pipe diameters and cavity gaps are given in Table 3.9. It can be observed that, for all the fixed cavity space cases, the larger pipe diameter will lead to more useful water heat gains. It is caused by the reduced friction loss that results in better water circulation and heat transfer, in particular at the narrow cavity space of 15mm(C15), enlarging the pipe diameter from 10mm(D10) to 15mm(D15) results in an obvious rise of water heat gain by 25%. With the pipe diameter varied from D15 to D20, the increase in useful water heat gain is around 7.3% for both cases with C15 and C20. Similarly, at larger cavity spaces of C20 and C25, the enlargement of pipe diameter from D20 to D25 results in about 2.5% improvement in useful water heat gains. Moreover, a tiny increase of 0.86% is observed with the pipe diameter increases from D25 to D30. It is indicated that the impact of pipe sizing is more evident at the smaller diameter levels.

Table 3.9 Variation in water heat gains with different cavity space and pipe diameter

(kW)

Cavity space/mm	C15			C20			C25		
Pipe size/mm	D10	D15	D20	D15	D20	D25	D20	D25	D30
Jan.	10.42	13.26	14.29	13.26	14.30	14.69	14.30	14.69	14.83
Feb.	7.02	9.42	10.30	9.41	10.30	10.64	10.30	10.63	10.77
Mar.	5.51	7.40	8.08	7.38	8.06	8.32	8.03	8.29	8.39
Apr.	3.21	4.20	4.57	4.13	4.51	4.65	4.44	4.59	4.64
May	3.95	5.01	5.38	4.94	5.32	5.46	5.26	5.40	5.45
Jun.	2.89	3.70	3.98	3.62	3.89	3.99	3.80	3.90	3.93
Jul.	2.44	3.18	3.43	3.08	3.32	3.41	3.21	3.30	3.32
Aug.	3.02	3.81	4.08	3.70	3.97	4.07	3.86	3.95	3.98
Sep.	3.81	4.67	4.99	4.58	4.89	5.01	4.79	4.91	4.95
Oct.	8.42	9.88	10.41	9.79	10.32	10.52	10.24	10.43	10.49
Nov.	11.86	14.20	15.04	14.18	15.03	15.35	15.02	15.33	15.44
Dec.	12.32	14.96	15.92	14.93	15.91	16.27	15.89	16.24	16.37
Sum	74.86	93.71	100.48	92.99	99.83	102.37	99.15	101.67	102.54

For the cases of C20 and C25, the local friction loss (i.e. fitting loss at heat exchanger) of D25 should be larger than D20 because the pipe size of 25mm is larger than the connected 20mm inner pipe diameter of the heat exchanger, while the friction loss along the flow path is smaller because of its larger diameter. The larger water heat gain under D25 implies that the impact of reduced friction loss along the flow path is more significant than the additional fitting loss at the heat exchanger.

At the same pipe size of D20, slightly higher water heat gain is obtained at C15 than C20 or C25; this is consistent with our previous findings. On the other hand, the larger the pipe diameter, the larger the water heat gain (so better operating performance). But from the system construction or embedded energy point of view, the increase in material use and system load should also be considered in the real case. Considering the insignificant improvement of 0.9% with the enlargement of pipe size from 25mm to 30mm, a connecting-pipe diameter of 20~25mm with the corresponding cavity space of 15~20mm is recommended for this buoyant driven window case.

The summary of yearly average thermal efficiency and room heat gain is given in Table 3.10. The minimum efficiency of 10.18% refers to the unfavorable case of cavity space 15mm and pipe diameter 10mm. At our recommended cavity thickness of 15~20mm and pipe diameter of 20~25mm, the thermal efficiency reaches 13.58% to 13.93%. Further increase of cavity and pipe size has no significant contribution to the thermal performance. Along with the increased water heat gain, the room heat gain is only slightly reduced by around 1%. It is indicated that the self-regulating buoyant-driven flow design has insignificant effect on the heat transmission through the window system. The improvement in useful water heat gain is mainly attributed to the increased water flow rate and the improved heat transfer coefficient at the heat exchanger.

Table 3.10 Thermal efficiency and room heat gain through the window system

Cavity space/mm	C15			C20			C25		
Pipe diameter/mm	D10	D15	D20	D15	D20	D25	D20	D25	D30
Thermal efficiency/%	10.18	12.75	13.67	12.65	13.58	13.93	13.49	13.83	13.95
Room heat gain/kW	214.0	212.3	210.7	207.3	205.4	204.1	200.3	199.1	198.2

Temperature and velocity variations during the typical summer (2nd ~ 8th Sep.) and winter (8th ~ 14th Jan.) weeks are used for illustration. Mass flow rates of water circulation in typical weeks are shown in Fig. 3.12, with C means cavity space and D stands for pipe diameter. For both summer and winter seasons, the mass flow rate increases with the enlargement of pipe size. And a larger increase is observed at the smaller pipe diameter levels. The minimum mass flow rate is obtained at C15 and D10, while the maximum one is at C25 and D30. At C25, when the pipe diameter increases from D25 to D30, the increase in mass flow rate is negligibly small. Consequently, as discussed above, insignificant difference in water heat gain is observed.

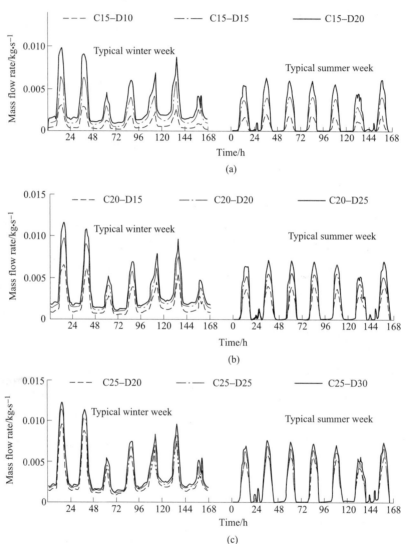

Fig. 3.12　Mass flow rates of circulation water with different cavity spaces
(a)15mm;(b)20mm;(c)25mm

The varied heat transfer coefficients at the heat exchanger are shown in Fig. 3.13. With the larger pipe size and faster water flow, higher heat transfer coefficient is obtained. Its variation is consistent with the curve of mass flow rate—the largest improvement is obtained with the change from D10 to D15, at the fixed C15. The increase in average heat transfer coefficient is around 11.5%. Similarly, at C25, the increase of pipe size from D20 to D30 has almost no contribution to the improvement of heat transfer coefficient.

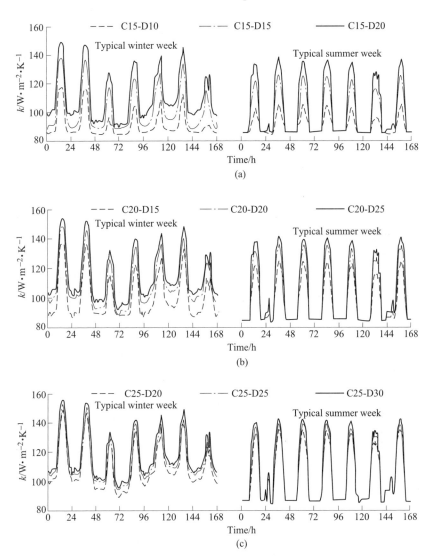

Fig. 3.13 Variation of heat transfer coefficient at the heat exchanger with cavity spaces
(a) 15mm; (b) 20mm; (c) 30mm

With the impacts on heat transfer as described, the liquid temperature distribution in the cavity is also affected. The temperature increases of circulating water in the cavity are shown in Fig. 3.14, with temperature abbreviated as 'Temp' for short. It can be observed that the trends of water temperature and flow rate changes are in the opposite direction.

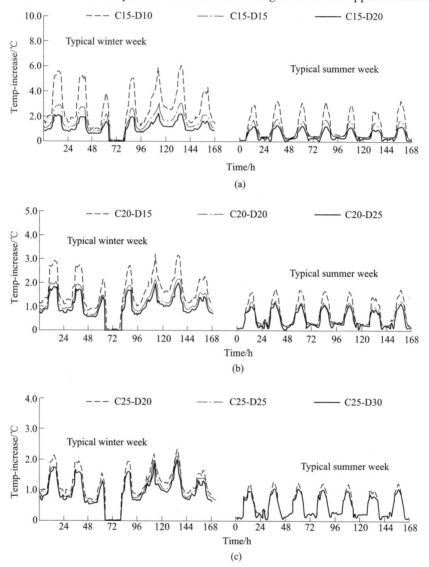

Fig. 3.14 Water temperature increase in the chamber with cavity spaces
(a) 15mm; (b) 20mm; (c) 25mm

The larger pipe size results in smaller temperature rise. The maximum temperature increases are in the ranges of 4~6℃ and 3~4℃ in the winter and summer seasons respectively, corresponding to the minimum flow velocity at C15-D10. At the smaller pipe size and larger friction loss situation, the return water from the heat exchanger is relatively low in temperature after the thermal release to the feed water stream. Further, because of the slow motion and low initial temperature at the window cavity, a larger temperature rise is achievable at longer retention time through both solar radiant and thermal convective heat absorption. This opposed effect on mass flow rate and temperature changes explains the very small variation in room heat gain across the seasons.

3.2.3 Thermal Performance Evaluation with Variation of Window Dimension

In this section, the influence of window dimension was studied and the water layer of 20mm determined in the previous section was used. For the optimized dimensions [1.2mm(H)×0.8mm(W)] decided based on the conceptual design, the total glazing area was 0.96m^2 and the glazing height to width ratio(GHTWR) was 1.5. However, this does not imply that the best length to width ratio is 1.5 when the glazing area is 0.96m^2. More studies should be carried out to make this clear. To verify this, numerical studies were conducted with glazing area of 1.6m^2, 0.96m^2 and 0.8m^2. Test cases with different height to width ratio are listed in Table 3.11, and the corresponding height to width ratio varied from 3.2 to 0.24. The variation of GHTWR influenced the system performance from two aspects. The first one was the variation of friction loss, which was related to the liquid flow and the heat transfer between the water layer and the glazing surfaces. And the second one was the heat exchanger effectiveness which was influenced by the variation in length of heat exchanger. Similarly, the useful water heat gain, the system thermal efficiency, the room heat exchanges through the window and the corresponding electricity saving caused by water flow window were compared.

Table 3.11 Test cases with three different glazing areas of 1.6m^2, 0.96m^2 and 0.8m^2 and five different height to width ratios

Area/m^2	1.6				
Length×Width	2.0×0.8	1.6×1.0	1.0×1.6	0.8×2.0	0.64×2.5
Length/Width	2.5	1.6	0.625	0.4	0.32
Area/m^2	0.96				
Length×Width	1.6×0.6	1.2×0.8	0.96×1.0	0.64×1.5	0.48×2.0
Length/Width	2.667	1.500	0.960	0.427	0.240
Area/m^2	0.8				
Length×Width	1.6×0.5	1.0×0.8	0.8×1.0	0.64×1.2	0.5×1.6
Length/Width	3.2	1.25	0.8	0.533	0.3125

The variations of monthly average thermal efficiency with different glazing areas are shown in Fig. 3.15 to Fig. 3.17. The increase in thermal efficiency with a decrease of length to width ratio could be observed for all the three cases. However, for the case with glazing area of 1.6m², a decrease of length to width ratio from 0.4 to 0.32 had very small contribution to the improvement of system thermal efficiency, but the increases from Jun. to Sep. were larger than the remaining eight months because of the higher ambient temperature during this period. For the remaining two cases, the decrease of length to width ratio to 0.24 and 0.3125 respectively even resulted in lowered system thermal efficiency, and the decreases in thermal efficiency from Jun. to Sep. were smaller as compared to the other eight months because of the higher outdoor air temperature. Besides, the effective glazing area was also different for windows with different length to width ratio because the heat exchanger was sealed in the window frame in a compact design of liquid flow window. The useful water heat gain and air-conditioning load caused by unit window area should also be different from that coming from unit glazing area.

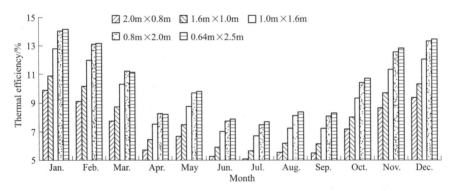

Fig. 3.15 Variation of thermal efficiency with glazing area of 1.6m²

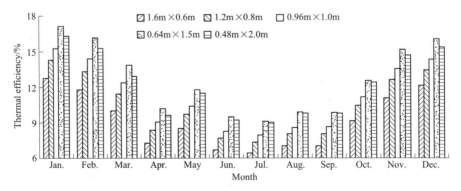

Fig. 3.16 Variation of thermal efficiency with glazing area of 0.96m²

It should be noted that with the same glazing area, the effective window area actually

3.2 Performance Evaluation with the Variation of Schematic Parameters

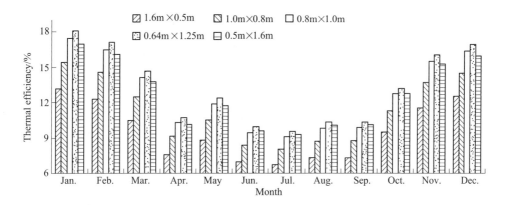

Fig. 3.17 Variation of thermal efficiency with glazing area of 0.8m²

changes with different glazing height to width ratio. Because of the presence of the heat exchanger which is sealed in the window frame in a compact manner. Thus, the useful water heat gain and thermal transmission per unit window area are smaller than those per unit glazing area. With a 0.1m vertical space reserved for heat exchanger installation, the results of useful water heat gain and thermal transmission normalized by the window area are given in Table 3.12. It can be seen that the water heat gain per unit window area decreases with the reduction in GHTWR. Taking the case with 0.96m² glazing area as an example, the reduction of this ratio from 2.667(1.6m×0.6m) to 1.5(1.2m×0.8m) contributes to an increase in water heat gains by 11.21%, but it can be 26.14% by decreasing the GHTWR from 2.667 to 0.427.

Also from Table 3.12, it can be seen that the room heat gain decreases with an increase in water heat gain; the corresponding room heat loss increases though the variation is small. An interesting point is that both water heat gain and room heat gain through the window turn to decrease with a further reduction in GHTWRs for instance at 0.3125, 0.24 and 0.32 for glazing areas of 0.8m², 0.96m² and 1.6m² respectively. Because at larger window width, the friction loss along the lengthy flow path becomes dominating and the water circulation decreases greatly, so will be the thermal extraction. But at the same time, the increase in the length of the heat exchanger improves its heat transfer effectiveness, leading to a drop in leaving water temperature. Consequently, the low water temperature profile in window cavity lowers the room heat gain. Overall speaking, a GHTWR around 0.4 can be appropriate for good system performance.

The saving in electricity caused by useful water heat gain, the electricity consumption of air-conditioning system, and the overall electricity saving caused by unit area of

Table 3.12 Normalized water heat gains and thermal transmissions
with different glazing areas   (kW·h/m^2)

Glazing Area/m^2	0.8				
GHTWR	3.2	1.25	0.8	0.533	0.3125
Window area/m^2	0.85	0.88	0.9	0.888	0.96
Water heat gains	83.08	95.16	104.97	110.78	97.73
Room heat gains	267.76	245.09	238.64	234.11	204.17
Room heat losses	−58.80	−56.80	−56.38	−59.59	−58.81
Glazing Area/m^2	0.96				
GHTWR	2.667	1.500	0.960	0.427	0.240
Window area/m^2	1.020	1.040	1.060	1.110	1.160
Water heat gains	80.03	89.00	93.60	100.95	93.23
Room heat gains	269.34	258.41	246.92	226.83	205.82
Room heat losses	−53.62	−54.93	−56.36	−56.58	−57.39
Glazing Area/m^2	1.6				
GHTWR	2.5	1.6	0.625	0.4	0.32
Window area/m^2	1.68	1.7	1.76	1.8	1.85
Water heat gains	62.99	69.43	78.59	84.99	83.98
Room heat gains	278.75	273.29	252.24	245.34	232.25
Room heat losses	−53.49	−54.13	−54.68	−54.72	−54.72

window are listed in Table 3.13. Similarly, the COP of air-conditioning system for cooling and heating were taken as 3.5 and 4.5 respectively, and the efficiency of electrical water heater was assumed to be 99%. The electricity consumption caused by the installation of water flow window could be reduced greatly because of the electricity saving in water heating. And the amount of overall electricity saving could be even larger than the electricity consumption of the air-conditioning system when the height to width ratio was reduced to the minimum value of 0.4 at the glazing area of 1.6m^2. As for the remaining two cases with smaller glazing area, the contribution in electricity saving caused by the reduction in height to width ratio was more evident. The electricity saving of water heater was larger than the energy consumption of air-conditioning system by 3.86kW·h and 13.47kW·h when the length to width ratio was reduced to 1.5 and 1.25. These were for windows with glazing area of 0.96m^2 and 0.8m^2, and the corresponding window dimensions were 1.2m(H)×0.8m(W) and 1.2m(H)×0.8m(W), respectively.

Table 3.13 Variation of normalized electricity saving with GHTWR at three different glazing areas ($kW \cdot h/m^2$)

Glazing Area/m^2	0.8			
GHTWR	3.2	1.25	0.8	0.533
Electricity saving in water heating	83.92	96.12	106.03	111.90
Electricity consumption in space cooling	76.50	70.03	68.18	66.89
Electricity consumption in space heating	13.07	12.62	12.53	13.24
Overall electricity saving	-5.65	13.47	25.32	31.77
Glazing Area/m^2	0.96			
GHTWR	2.667	1.500	0.960	0.427
Electricity saving in water heating	80.84	89.90	94.55	101.97
Electricity consumption in space cooling	76.95	73.83	70.55	64.81
Electricity consumption in space heating	11.92	12.21	12.52	12.57
Overall electricity consumption	-8.03	3.86	11.48	24.59
Glazing Area/m^2	1.6			
GHTWR	2.5	1.6	0.625	0.4
Electricity saving in water heating	63.63	70.13	79.38	85.84
Electricity consumption in space cooling	79.64	78.08	72.07	70.10
Electricity consumption in space heating	11.89	12.03	12.37	12.16
Overall electricity consumption	-27.90	-19.98	-5.06	3.59

3.3 Summary

The effect of several configuration design parameters on the water flow window performance has been successfully analyzed numerically with a self-developed FORTRAN program. The impact of liquid layer thickness was first analyzed, then the thermal performance under different pipe-work designs. Following this, window dimension optimization from the height to width ratio viewpoint was completed. The major conclusions are drawn below:

(1) Firstly, the water layer thickness has shown an impact on the system energy performance, since the thermal resistance is reduced with a thinner water layer. The water temperature increases faster with a smaller amount of water in the cavity. The decrease of water layer thickness from 30mm to 20mm has more evident influence on the system overall performance. Further reduction of water layer thickness from 15mm to 10mm deteriorates the system performance because of the increase in air-conditioning load. And

the energy saving potential of the window system with 20mm water layer is the largest, considering the overall electricity saving. A water layer of 15~20mm can be desirable for achieving good energy performance in terms of electricity saving based on the glazing dimensions of 1.2m(H)×0.8m(W).

(2) By eliminating the distribution headers, the water circulation can be improved and the temperature distribution at the inner glazing surface is more uniform. The improvement in yearly water heat gain is about 5%, with a corresponding minor drop in room heat gain.

(3) For a window system without distribution headers, its thermal performance is also affected by the connecting-pipe size, especially for the smaller diameter cases. The increase in pipe diameter results in reduced friction loss and consequently, the improved water circulation and heat transfer coefficient. As a result, better heat transfer at the heat exchanger is realized. Promising thermal performance can be achieved with the use of proper return-pipe size; the recommended range is 20~25mm through which comparative high thermal efficiency of 13.58% to 13.93% can be realized.

(4) For a window at fixed glazing area, a reduction in length to width ratio can help to improve the system performance because of the improvement in heat exchanger effectiveness. The minimum height to width ratio is around 0.4 for good performance in most of the studied cases. The electricity saving from useful water heat gain may be even larger than the electricity consumption of air-conditioning system if a proper height to width ratio is adopted.

Overall speaking, the water flow window without distribution headers is an energy-efficient and economical design. Meanwhile, a proper determination of cavity and connecting-pipe size, as well as dimension is also of high importance.

4 Performance Evaluation with PCM Thermal Storage for Residential and Office Buildings Hot Water Demand Patterns

Phase Change Material (PCM) has larger heat storage capacity comparing to the traditional gas and liquid materials. Phase transition takes place at almost constant temperature and latent heat energy is stored or released during this process. Its large heat storage capacity with small volume is very attractive. With incorporation of PCM in the water heating system, hot water supply period can be elongated effectively. And thus the product is widely used for the purposes of thermal storage and peak load shifting.

PCM is widely used in solar thermal energy storage for its high heat storage density through latent heat exchange. The storage capacity of PCM per unit volume can be 5 to 14 times of the sensible heat storage media, like water. Abhat categorized the low temperature latent heat storage materials according to their properties. Sharma et al. , reviewed the thermal energy storage properties of PCM; The great application potential in solar engineering systems was illustrated and the importance of selecting PCM in accordance with the melting temperature was highlighted. The supply period of hot water could be extended effectively because of the additional thermal storage. Medrano et al. evaluated the application of PCM in commercial heat exchangers; It is reported that higher efficiency could be achieved. On the other hand, the charging and discharging performance of the heat exchanger can be improved by using combined PCM with different phase transition temperatures. The results were found affected by the temperature span as well as the arrangement of combined PCMs.

As a key component in thermal storage, the performance and heat transfer mechanism of various types of heat exchangers have been extensively studied. Many experimental, analytical and simulation-based research efforts have been on studying the PCM melting and solidification mechanism. In their extensive reviews, Agyenim et al. pointed out that for double-pipe heat exchanger, the heat transfer in the PCM layer is primarily dominated by thermal conduction, whereas the temperature variation along the axial direction is generally small as compared to that along the radial direction. The numerical analysis of such problem is complicated by the movement of the solid-liquid boundary. This can be well overcome by using the enthalpy method. Some research studies with PCM encapsulation in simple structures like pipe or rectangular cavity were carried out with the CFD tool 'FLUENT', in which the enthalpy model was adopted as well. Very good accuracy was demonstrated through model validation. In other studies, enthalpy-model based codes were also developed and embedded in the TRNSYS system simulation or MATLAB for system performance prediction.

The focus of this Chapter is on the other major system component, the heat exchanger. The double-pipes heat exchanger used in this system is of simple structure and low cost, but of low efficiency and thus there are rooms for improvement in heat transfer performance. Despite the demonstrated attractive energy saving and cost payback potential, mismatch in hot water production and demand may exist in different building types. In residential buildings, the hot water demand is relatively high in the early morning and evening when the solar radiation is weak. Thermal storage and load shifting is then advantageous to cater for the situation. The potential application of PCM is considered in this study as a mechanism providing load shifting and heat transfer enhancement. The successful numerical validation in Chapter 2 indicates that the self-developed computer program is able to predict the flow and heat transfer performance of liquid flow window with good accuracy. In this chapter, this validated program will be used to assess the effectiveness of system performance improvement schemes.

The PCM can be placed at different positions in the heat exchanger. In the present study, the performances of two cases were assessed. In Case 1, the PCM layer was located in the annular space between the hot fluid and the cold water; and in Case 2, the PCM layer was added to the outermost layer of the heat exchanger. The arrangement of hot water, cold water and PCM in the two designs and the size of the heat exchanger are shown in Fig. 4.1. In the figure, H means hot water, C means cold water and P means PCM layer. Thus the original double-pipe heat exchanger was changed into a triple-pipes heat exchanger. The design of Case 1 was named as HPC, and that of Case 2 was named as HCP according to the position of different materials.

The PCM adopted in this numerical study was RT35HC, which is a kind of pure PCM capable of storing and releasing large amount of thermal energy at almost constant temperature. It was chemically inert and without super cooling effect. Its lifetime was long with a stable performance over several ten thousands of working cycles. Some of its thermo-physical properties are listed in Table 4.1. The thickness of PCM layer used in this study was 0.01m.

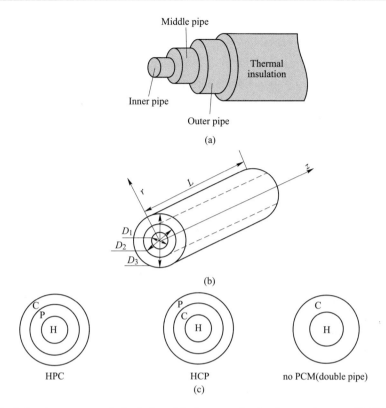

Fig. 4.1 Arrangement of physical model of concentric-pipe heat exchanger with PCM
(a) 'Cut-away' view of insulated heat exchanger; (b) Model configuration and coordinate;
(c) Layers of heat-carrying materials in pipe

Table 4.1 Thermo-physical properties of RT35HC

Melting area/℃	34~36(main peak:35)
Congealing area/℃	36~34(main peak:35)
Heat storage capacity (±7.5%)/kJ · kg^{-1}	240
Specific heat capacity/kJ · (kg · K)$^{-1}$	2
Density(solid)/kg · L^{-1}	0.77
Density(liquid)/kg · L^{-1}	0.67
Heat conductivity(both phases)/W · (m · K)$^{-1}$	0.2
Volume expansion/%	12
Max. operation temperature/℃	70

4.1 Numerical Models Development

In this numerical study, heat transfer in the PCM layer was solved with the enthalpy method. To simplify the PCM heat transfer problem, the following assumptions were made in the numerical model formulation process:

(1) The thermal properties of the heat transfer fluid and the PCM layer in the heat exchanger are temperature independent, but the thermal properties of PCM are different in the solid and liquid states.

(2) Convective heat transfer within the PCM layer is negligible; only isothermal phase change is considered during the melting-solidification process.

(3) The outer surface and the two end surfaces of the shell are well-insulated and thus behave as adiabatic system boundaries.

(4) Radial conduction heat transfer in water is negligible.

Numerical models for heat transfer calculation at the heat exchanger are thus developed.

4.1.1 Mathematical Models for Case 1

Consider a PCM-incorporated heat exchanger of length L and with PCM added at the annular space between the hot and cold flow streams as indicated in Fig. 4.1(c). The developed mathematical models are given below:

(1) Based on the above assumptions, the energy equation of the PCM layer in 2-D cylindrical coordinate [Shown in Fig. 4.1(a)] is given by

$$\frac{\partial h_p}{\partial t} = \frac{1}{r}\frac{\partial}{\partial r}\left(r\alpha_p \frac{\partial h_p}{\partial r}\right) + \alpha_p \frac{\partial^2 h_p}{\partial z^2} - \rho_p \lambda \frac{\partial f}{\partial t} \quad (4-1)$$

Where α_p ——the thermal diffusivity of PCM in m^2/s;

ρ_p ——the density of PCM in kg/m^3;

λ ——the latent heat of phase transition in J/kg;

$h_p(T)$ ——the sensible enthalpy in J/kg, given by

$$h_p(T) = \int_{T_m}^{T_p} \rho_p c_p dT \quad (4-2)$$

Where c_p ——the specific heat of PCM in $J/(kg \cdot K)$;

T_p ——the temperature of PCM layer in $°C$;

T_m ——the phase transition temperature in $°C$.

(2) The mathematical model of hot water for Case 1 is given by

$$\rho_f c_f \left(\frac{\partial T_h}{\partial T} + u\frac{\partial T_h}{\partial z}\right) = \frac{4h_h}{D_1}(T_p - T_h) + k_f \frac{\partial^2 T_h}{\partial z^2} \quad (4-3)$$

(3) And that of cold water is written as

$$\rho_f c_f \left(\frac{\partial T_c}{\partial T} + u\frac{\partial T_c}{\partial z}\right) = \frac{4h_c \cdot D_2}{D_3^2 - D_2^2}(T_p - T_c) + k_f \frac{\partial^2 T_c}{\partial z^2} \quad (4-4)$$

(4) The boundary conditions for PCM layer is given by

$$h(T_p - T_h) = k_p \frac{\partial T_p}{\partial r}, \quad r = r_{p1} \tag{4-5a}$$

$$h(T_p - T_c) = -k_p \frac{\partial T_p}{\partial r}, \quad r = r_{p2} \tag{4-5b}$$

$$\frac{\partial T_p}{\partial z} = 0, \quad z=0 \text{ and } z = L \tag{4-6}$$

(5) Boundary conditions for hot and cold water inlet is given by

$$T_h = T_{hin} \quad (z = 0) \tag{4-7}$$

$$T_c = T_{cin} \quad (z = 0) \tag{4-8}$$

Where T_{hin} ——the inlet temperature of hot water in ℃;

T_{cin} ——the inlet temperature of cold water in ℃.

(6) Boundary conditions for the hot and cold water outlets are determined by assuming that the flow is thermal fully-developed at the outlet.

$$\frac{\partial T_h}{\partial z} = 0 \quad (z = L) \tag{4-9}$$

$$\frac{\partial T_c}{\partial z} = 0 \quad (z = L) \tag{4-10}$$

4.1.2 Mathematical Models for Case 2

The arrangement of PCM-incorporated heat exchanger with PCM added at the outermost layer is indicated in Fig. 4.1(c). The developed mathematical models are as follows:

(1) The mathematical model of the PCM layer is the same as in Case 1.

(2) The energy balance equation of the hot water stream is given by

$$\rho_f c_f \left(\frac{\partial T_h}{\partial t} + u_h \frac{\partial T_h}{\partial z} \right) = \frac{4h_h}{D_1}(T_c - T_h) + k_f \frac{\partial^2 T_h}{\partial z^2} \tag{4-11}$$

Where ρ_f ——the density of water in kg/m^3;

c_f ——the specific heat of water in J/(kg·K);

k_f ——the heat conductivity of water in W/(m·K);

u_h ——the flow velocity of hot water in m/s;

h_h ——the convective heat transfer coefficient of hot water in W/(m^2·K);

T_c, T_h ——the temperatures of cold and hot water in ℃;

D_1 ——the diameter of the innermost pipe in m shown in Fig. 6.1.

(3) The energy balance equation for cold water is written as

$$\rho_f c_f \left(\frac{\partial T_c}{\partial T} + u_c \frac{\partial T_c}{\partial z} \right) = \frac{4h_c D_1}{D_2^2 - D_1^2}(T_h - T_c) + \frac{4h_c D_2}{D_2^2 - D_1^2}(T_p - T_c) + k_f \frac{\partial^2 T_C}{\partial z^2}$$

(4-12)

Where u_c ——the flow velocity of cold water in m/s;

h_c ——the convective heat transfer coefficient of cold water in W/(m² · K);

D_2 ——the outer diameter of the annular space in m, as observed in Fig. 6.1.

(4) The boundary conditions for the PCM layer at $r = r_{p_1}$ and $r = r_{p_2}$ are given by

$$h(T_p - T_c) = k_p \frac{\partial T_p}{\partial r}, \quad r = r_{p_1} \tag{4-13a}$$

$$\frac{\partial T_p}{\partial r} = 0, \quad r = r_{p_2} \tag{4-13b}$$

Where k_p ——the thermal conductivity of PCM in W/(m · K);

r_{p_1} ——the outer diameter of inner tube in m;

r_{p_2} ——the inner diameter of the outer tube in m.

$$\frac{\partial T_p}{\partial z} = 0 \quad (z = 0, \text{ and } z = L) \tag{4-14}$$

The boundary conditions of hot and cold water are the same as in Case 1.

4.2 Model Validation

A validation exercise was conducted to prove that the above-mentioned model could be used to model the heat transfer mechanism of the PCM layer with adequate accuracy. It was conducted based on the published experimental data of Lacroix. The study focused on the heat transfer mechanism of a shell and tube heat exchanger with hot water flowed in the inner tube and PCM filling the shell side.

The heat exchanger was 1m long. The inner and outer diameters of the inside tube was 0.0127m and 0.0158m, respectively, and the thickness of PCM layer was 0.005m. The experimental configuration is as shown in Fig. 4.2.

N-octadecane was used as the phase change material and its thermo-physical properties are given in Table 4.2. The experiment was carried out in three different conditions. The hot water velocity was maintained at 0.0315kg/s(corresponding to 0.24m/s) for all the three cases, but the inlet temperature of hot water was 5℃, 10℃ and 20℃, higher than the fusion temperature, respectively. The initial temperature of PCM layer was set at 9.5℃.

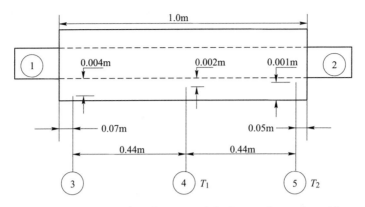

Fig. 4.2 Experimental configuration of the heat exchanger in article
1—Thermocouples; 2—Inside the heat transfer fluid; 3—Thermocouples; 4—Inside paraffine; 5—Wax

Table 4.2 Thermo-physical properties of N-octadecane

Fusion temperature/℃		27.5
Latent heat of fusion/kJ · kg^{-1}		243.5
Thermal conductivity/W · (m · k)$^{-1}$	Liquid	0.148
	Solid	0.358
Thermal diffusivities/m^2 · s^{-1}	Liquid	8.64×10^{-8}
	Solid	2.14×10^{-7}
Density/kg · m^{-3}	Liquid	782
	Solid	820
Heat capacity/J · (kg · k)$^{-1}$	Liquid	2190
	Solid	1944

The mathematical model used for heat transfer calculation of the PCM layer is the same as Case 1. So the mathematical model of the hot water is given by

$$\rho_f c_f \left(\frac{\partial T_h}{\partial t} + u_h \frac{\partial T_h}{\partial z} \right) = \frac{4h_h}{D_1}(T_p - T_h) + k_f \frac{\partial^2 T_h}{\partial z^2} \qquad (4\text{-}15)$$

Where ρ_f ——the density of hot water in kg/m^3;

c_f ——the specific heat of hot water in J/(kg · K);

k_f ——the heat conductivity of hot water in W/(m · K);

u_h ——the flow velocity of hot water in m/s;

h_h ——the convective heat transfer coefficient of hot water in W/(m^2 · K);

T_h, T_p ——the temperatures of hot water and PCM at the corresponding node points in ℃;

D_1 ——the diameter of the inner tube with hot water flow in m.

The boundary conditions for hot water are the same as in Case1. The boundary conditions for PCM layer are given by

$$h(T_p - T_h) = k_p \frac{\partial T_p}{\partial r}, r = r_{p1} \qquad (4\text{-}16)$$

$$\frac{\partial T_p}{\partial r} = 0, r = r_{p2} \qquad (4\text{-}17)$$

$$\frac{\partial T_p}{\partial z} = 0, z = 0 \text{ and } z = L \qquad (4\text{-}18)$$

Where r_{p1}——the outer diameter of inner tube in m;

r_{p2}——the inner diameter of the outer tube in m.

The validation study was completed by taking the inlet temperature of hot water as 37.5℃ (10℃ higher than the phase transition temperature) and 47.5℃ (20℃ higher than the phase transition temperature). In the simulation, the thickness of the inner tube was considered but its thermal resistance was neglected. The result comparisons of T_1 and T_2 in the PCM layer from simulation and experiment are shown in Fig. 4.3 and Fig. 4.4. Temperatures obtained from the simulation agreed well with the published ex-perimental

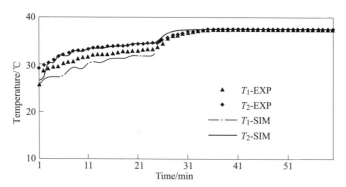

Fig. 4.3 Comparison of simulation and Lacroix's experiment results for $\Delta T = 10℃$

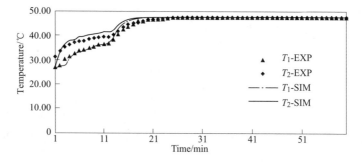

Fig. 4.4 Comparison of simulation and Lacroix's experiment results for $\Delta T = 20℃$

data. So this model could be used for predicting the heat transfer conditions in the PCM layer.

4.3 Numerical Study on Thermal Performance of Liquid Flow Window with PCM Heat Exchanger

Simulation studies were carried out to assess the performance of the two alternative designs with PCM as compared to the original one to be used in residential and office buildings. It was conducted during typical summer(2nd ~ 8th Sep.) and winter(8th ~ 14th Jan.) weeks of Hong Kong with the TMY weather data. Temperature of indoor environment was preset at 21℃ and 25℃ respectively in winter and summer weeks. The hot water demands of the residential and office buildings are different. The water demand patterns of residential and office buildings by percentage are given in Fig. 4.5. In residential building, hot water supply is required from 7am and the peak water demand in the morning occurs at 8 am. The demand decreases after that and reaches another high point at noon, and then the water demand decreases again gradually. The peak hot water demand in the evening appears between 6pm and 8pm during off work hours. For office buildings, there is no hot water demand before 8 am and after 6pm. The hot water demands at 8am and 5pm are also smaller than the other hours with large difference. But the demand of hot water in the remaining hours is almost stable with the peak demand appears at noon and 4pm.

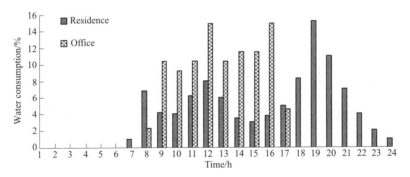

Fig. 4.5 Water demand patterns of residential and office buildings by percentage

In this study, the hourly water flow rates were derived from the above hot water demand patterns. The cold water flow rate used in the numerical study for residential and office buildings are listed in Table 4.3. The daily hot water supply from a single piece of window for residential application is 78.9L, and this is 80.28L for office building. System performances were compared from the aspects of water heat gain and thermal transmission through the window system.

Table 4.3 Cold water flow rate for residential and office buildings determined by demand patterns (mL/min)

Time/h	Residence	Office
1~6	0	0
7	13	0
8	89	31
9	55	138
10	53	123
11	82	138
12	105	200
13	79	138
14	46	154
15	39	154
16	50	200
17	66	62
18	109	0
19	200	0
20	145	0
21	92	0
22	53	0
23	26	0
24	13	0

4.3.1 Thermal Performance for Residential Building Application

The variations of solar radiation on the tested window surface during typical summer and winter weeks are shown in Fig. 4.6, they were obtained from the TMY hourly weather data set of China available from the US-DOE website. The water heat gain and indoor air-conditioning load were affected by the solar radiation level greatly. For most of the time, the solar radiation during typical summer week is stronger than that of typical winter week.

Daily water heat gains of the three cases during typical summer and winter weeks for residential use are shown in Fig. 4.7 and Fig. 4.8 respectively. On each day, it includes the sum of water heat gain from 7 am to 12pm during the period with hot water demand. The difference in water heat gain between Case 2 and Case 3 was very tiny. But the water heat gain of Case 1 with PCM located in the annular space was much smaller than the

4.3 Numerical Study on Thermal Performance of Liquid Flow Window with PCM Heat Exchanger

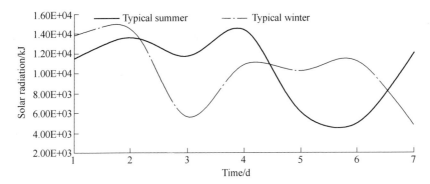

Fig. 4.6 Solar radiation incident on window surface in typical summer and winter weeks
(Note: typical summer week period: 2nd ~ 8th Sep. ; Typical winter week period: 8th ~ 14th Jan.)

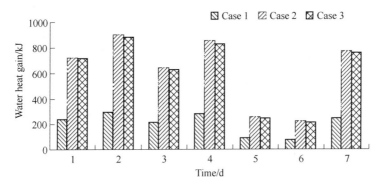

Fig. 4.7 Water heat gain of residential building during typical summer week

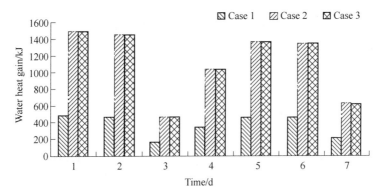

Fig. 4.8 Water heat gain of residential building during typical winter week

other two cases for both summer and winter seasons. The difference is in the range of 66% ~ 67% because firstly the cold water flowing in the annular space, which was next to the flow path of hot water in Case 2 and Case 3, and thus it could be heated directly. Another reason was that the convective heat transfer took place in both hot and cold water streams around the same pipe wall, and thus cold water could be heated up more effectively.

For Case 1, because of the addition of PCM in the annular space, the thermal resistance was increased and a large amount of thermal energy was absorbed during the melting process of PCM. Then the amount of thermal energy transferred to the cold water was very small. Thus the temperature increase of cold water was not evident. And consequently, the cold water heat gain was small in value.

The water heat gain of Case 2 was slightly larger than that of Case 3 when the solar radiation was strong because a part of the thermal energy absorbed by the cold water was extracted by the PCM layer for Case 2, and the thermal energy stored in the PCM layer would be released back to the cold water later on when the temperature of the PCM layer decreased down to the phase transition point. At the same time, heat transfer between hot and cold water in the heat exchanger could be further enhanced because of the thermal extraction of the PCM layer. The favorable water heat gains of Case 2 above Case 3 show the merit of PCM storage especially under strong solar radiation. On Day 5 and Day 6 of the typical summer week when the solar radiation was weak, the differences in water heat gains among the three cases were small, and a similar phenomenon could be observed from the results of Day 3 during the typical winter week. This indicated that the system could not benefit much from PCM thermal storage when the solar radiation was weak.

In typical summer week, the total water heat gains were 4347kJ and 4277kJ respectively for Case 2 and Case 3. Hence the water heat gain was increased by 1.63% in typical summer week by adding PCM to the outermost layer of the heat exchanger. While in typical winter week, they were 7783kJ and 7773kJ respectively, and the increase was only about 0.13%. However, more hot water was supplied during off-work hours. It was observed that during the off-work hours, 17.4% more thermal energy could be obtained in Case 2 than in Case 3 during the typical summer week. The gap was about 3% within the typical winter week. The lowered winter performance could be the consequence of weakened solar radiation and lower ambient temperature. It is indicated that the system benefits more from the addition of PCM in summer than in winter. The average increase in off-work hot water supply during typical summer and winter weeks was 159kJ, which is equivalent to 8291kJ per year. The total water heat gain during typical winter week was larger than that in typical summer week because the inlet temperature of cold feed water was low in the typical winter week. The average increase in water heat gain of the residential building during typical summer and winter weeks was about 0.88%, with an average cold feed water flow rate of 73mL/min.

However, the effects of heat storage and load shifting of PCM could not be observed in the previous two figures. The useful water heat gains during the first two days of the typical summer week are used here to illustrate its hourly variation, as shown in Fig. 4.9. There existed heat release for Case 2, and this happened after 6pm. The peak value of water heat gain appeared at 7pm, when the hot water demand was the largest for residential building. For the case without PCM, the peak water heat gain occurred at 5pm and was lower than Case 2 on the same day 3rd Sep. This larger water heat gain of Case 2 at 7pm was caused by load shifting as a result of the stored heat at the PCM layer. The peak water heat gain of Case 3 occurred at 5pm. Its magnitude was lower than that of Case 2. For Case 1, the maximum water heat gains also occurred at 7pm because of its thermal storage effect. This was consistent with the observation in Case 2. But the total amount of water heat gain remained small because of the large thermal resistance of the PCM layer located between the hot and cold water layers.

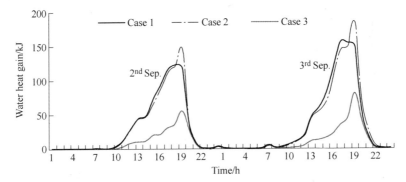

Fig. 4.9   Water heat gain of residential building on the first two days of typical summer week

Apart from the effect on water heat gain, thermal transmission through the window system was also affected by the design of heat exchanger because the return water temperature could be different. In the summer season, only room heat gain was affected. However, in the winter season, both room heat gain and heat loss were affected. The room heat exchanges through this 0.98m^2 window surface during typical summer and winter weeks are shown in Fig. 4.10 and Fig. 4.11.

In typical summer week, the room heat gain of Case 1 was larger than the other two cases. This was because the thermal energy extracted by the flowing liquid layer was not well released to the cold feed water when PCM was added to the annular space. Instead, a part of the PCM stored energy has been transferred back to the buoyant-driven water stream since the layer of PCM adjacent to the hot water stream was of the highest tem-

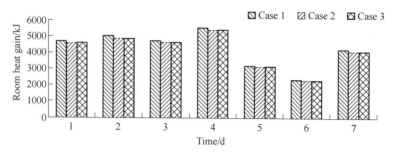

Fig. 4.10 Room heat gains through the window for residential building during typical summer week

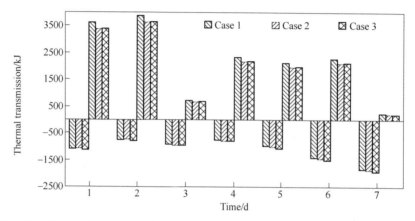

Fig. 4.11 Room heat exchanges through the window for residential building during typical winter week

perature. And this would cause an increase in glazing surface temperature. Room heat gains of Case 2 and Case 3 were 29115kJ and 29102kJ respectively with a difference of about 0.047%. The room heat gain of Case 2 was found lower than that of Case 3 during the daytime because of the thermal extraction of the PCM layer from the water stream, and consequently resulted in a slightly lower return water temperature. But the heat release of PCM during the night time leaded to an increase in return water temperature. These together brought about relatively stable and uniform daily solar transmissions in these two cases.

Similarly, in typical winter week, the room heat gain of Case 1 was larger than the other two cases. However, the room heat loss through the window varied in an inverse way. This was caused by the same reason of higher window surface temperature with PCM located in the annular space between hot and cold water streams. Heating was required when the indoor air temperature was lower than the set-point. A lower window surface temperature of Case 2 and Case 3, which was good for reduction in room heat gain, would cause an increase in room heat loss in the winter season. But the difference in room heat losses was much smaller than in room heat gains.

The quantity of thermal storage in the PCM layer is affected by the daytime thermal extraction of the cold water stream. Only limited amount of thermal energy could be stored in the PCM layer during the daytime since large amount of thermal energy was carried away by the cold feed water stream. In order to enhance hot water production during the off-work hours, the thermal storage performance was also assessed by assuming zero cold feed water flow during the daytime. And instead, the feed water flow was taken within the off-work hours from 6pm to 9pm.

The water heat gain during typical summer week in this case was compared to the results coming from the designated demand pattern of residential buildings as in Fig. 4.7. Comparison of water heat gains during typical summer weak is given in Table 4.4. By assuming the zero water flow during daytime, the water heat gain of Case 3 was affected greatly because of the sacrificing of thermal absorption during daytime when solar radiation was strong. It was found to have the least impact on the water heat gain of Case 1 because the thermal energy carried by the hot water was used to heat the PCM layer in the first place. After that, the heat energy could be released to the cold water stream when the temperature of the PCM layer was high enough. And thus the amount of thermal energy absorbed by the cold water was still of small amount. Under this condition, the performance of Case 2 was the best because of its thermal storage capability, but the weekly total cold water heat gain was still reduced by 43% as compared to the actual supply pattern.

Table 4.4 Comparison of total water heat gain during typical summer week

Total water heat gain/kJ	Case 1	Case 2	Case 3
Designated flow pattern	1441	4347	42767
Assumed off-work water supply	1054	2473	1887
Difference	387	1874	2390

The results were different when comparing only the water heat gain during off-work hours from 6pm to 9pm as given in Fig. 4.12. The water heat gain during off-work hours would increase by supplying water during off-work peak hours as compared to the design with actual flow pattern. It can be seen that under this circumstance the difference in water heat gains of Case 2 is the largest. With the assumed off-work water supply running mode, the water heat gains during the off-work hours of Case 2 at 2473kJ is greater than that of Case 3 at 1887kJ by 31.4%. This is more than the difference of 17.4% based on the whole-day running mode. Similarly, the difference in the typical winter week is

11.4%, which is also greater than the 3% difference of the whole-day running mode. The benefit of PCM incorporation is thus enlarged. There can be two reasons. Firstly, it can be caused by the heat release from the PCM layer to the incoming cold water stream. Secondly, the stationary cold water is heated to a fairly high temperature in the heat exchanger for all day long until 6pm. The average increase in water heat gains during off-work hours reaches 470kJ/week during typical summer and winter weeks. It is much larger than the whole day running mode. But the increase in water heat gain during off-work hours was achieved by sacrificing the large amount of water heat gain during daytime.

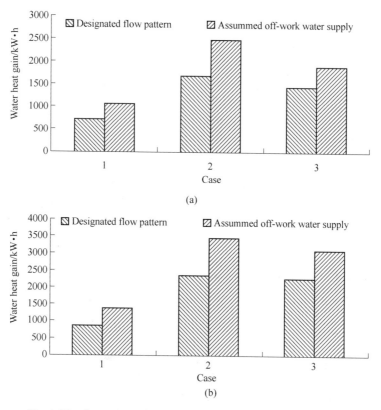

Fig. 4.12 Comparison of total water heat gains during off work hours
(a) Typical summer; (b) Winter weeks

And zero flow during daytime could also result in increase in room heat gains as given in Fig. 4.13, though the significance was considerably less than the effect of water heat gain. The increase of Case 2 was the largest as compared to the other cases as a result of thermal storage and release of the PCM layer, and the increase of Case 1 was of the smallest value. However, the increase in room heat gain during daytime was unlikely to result in significant increase in energy consumption for residential use. In the absence of occupancy during daytime, air-conditioning is generally not required.

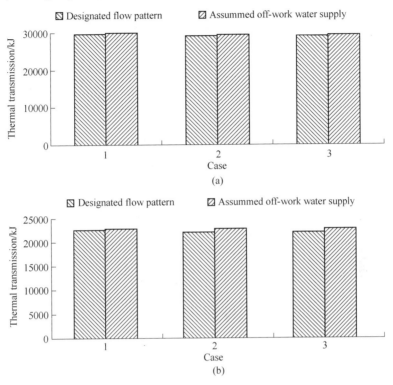

Fig. 4.13 Comparison of thermal transmission
(a) Typical summer; (b) Winter weeks

4.3.2 Thermal Performance for Office Building Application

For the application in office buildings, simulations were carried out under the same weather condition. The water heat gains and room heat exchange through the window during typical summer and winter weeks were also compared to study the impacts of PCM application. Variations of water heat gain during typical summer and winter weeks are given in Fig. 4.14 and Fig. 4.15. Similar to the results of residential buildings, water heat gain of Case 1 was the smallest during both typical summer and winter weeks. And the water heat gain of Case 2 remained to be the largest as compared to Case 3, though the difference was very small during typical winter week. The large amount of water heat gain of Case 2 in office building came from the heat release of stored thermal energy in the PCM layer to the cold feed water in the early morning.

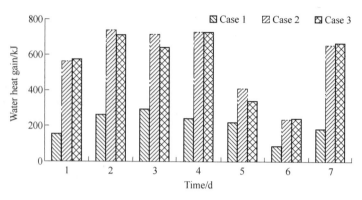

Fig. 4.14 Water heat gain of office building during typical summer week

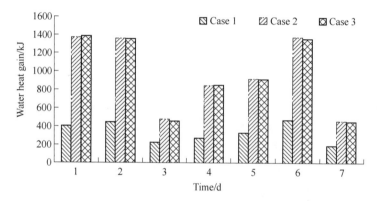

Fig. 4.15 Water heat gain of office building during typical winter week

In typical summer week, total water heat gain of Case 2 and Case 3 were about 2.8 and 2.7 times of Case 1. Water heat gain of Case 2 and Case 3 were 4045kJ and 3897kJ respectively with difference about 3.8%. This increase was slightly larger than the increase in residential buildings because that heat transfer between hot and cold water was better with larger cold feed water velocity in office use, and thus the thermal extraction of the PCM layer was also increased.

During typical winter week, performance of Case 2 and Case 3 remained to be better as compared to that of Case 1. But the water heat gain of Case 3 was still smaller than that of Case 2 with values of 6779kJ and 6808kJ respectively, corresponding to a difference of 0.44%. The difference was smaller than that of typical summer week as a result of the weak solar radiation. And this was consistent with the results in residential use. In both typical summer and winter weeks, differences in water heat gain of Case 2 and Case 3 for office application were always larger than that of residential use because of its relatively large hot water demand and consequently, the higher feed water velocity. This indicated that adding of PCM at the outermost layer of the heat exchanger was more advantageous for office use. The average increase in water heat gain of the office building during typical summer and winter weeks was about 2.13% with an average cold feed water flow rate of 134ml/min.

The thermal transmissions through the window during typical winter weeks were also compared. The variation in room heat exchanges through the window during typical summer and winter weeks are shown in Fig. 4.16 to Fig. 4.17. During typical summer week, Case 1 had the largest room heat gain, while the room heat gain of Case 2 was almost equal to that of Case 3; and it was similar to the phenomenon of residential use. The difference in room heat gain between Case 1 and the other cases could be as large as 655kJ, which was close to the difference in residential application during typical summer week.

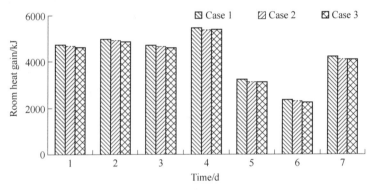

Fig. 4.16 Room heat gains of office building during typical summer week

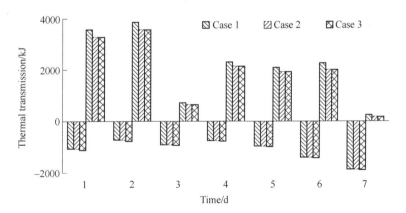

Fig. 4.17 Room heat exchanges of office building during typical winter week

During typical winter week, the room heat gains of Case 2 and Case 3 were smaller than that of Case 1. The reduction in cooling load from Case 1 to Case 2 and Case 3 were

1150kJ and 1156kJ, corresponding to differences of 8.32% and 8.37% respectively. The differences in room heat gain of Case 2 and Case 3 during typical summer and winter weeks were negligible with values of 39kJ and 6kJ. And this small increase in room heat loss of Case 2 as compared to Case 3 was also caused by the heat release of the PCM layer in the early morning. Similarly, the room heat losses of Case 2 and Case 3 during typical winter week were slightly larger than that of Case 1 as concluded from the simulation results of residential use. The increase in room heat loss from Case 1 to Case 2 was 114kJ. It was smaller than the corresponding increase of 360kJ in residential use. And the reason was that there was no thermal extraction of water stream after 6pm for office buildings, and the window surface temperature would not be further reduced. For residential use, the extraction of thermal energy by cold water during nighttime resulted in further reduction of the window surface temperature, which caused an increase in room heat loss. However, the larger amount of room heat losses of Case 2 and Case 3 as compared to Case 1 was favorable for office use by offsetting the cooling loads coming from occupant, lighting and equipment.

4.3.3 Thermal Performance with Variation of PCM Layer Thickness

Contributions to thermal extraction were different with variation in PCM layer thickness. In the present study, two more cases were brought into consideration with PCM layer of 12mm and 5mm for Case 2. System performance was compared considering the factors of water heat gain and thermal transmissions through the window. The comparisons of water heat gain for residential and office buildings are given in Table 4.5. For both residential and office buildings, the total water heat gain increased with the increase in PCM layer thickness during both typical summer and winter weeks. But the increase in typical summer week was more evident because of the higher ambient temperature and solar radiation level. And the increase of PCM layer from 5mm to 10mm contributed to the major increase in useful water heat gain during the typical summer week.

Table 4.5 Comparison of water heat gain with different PCM layer thickness

(kJ)

	Thickness/mm	12	10	5
Residence	Typical summer	4349	4347	4328
	Typical winter	7786	7783	7777
Office	Typical summer	4045	4045	4003
	Typical winter	6811	6808	6797

The comparisons of thermal transmission through the window are given in Table 4.6. The variation in PCM layer thickness had negligible small effect on the indoor heat gain and heat loss, with a difference of less than 0.02%. For residential use, the increase in PCM layer thickness resulted in a slightly decrease in indoor cooling loads in both typical summer and winter weeks because the thermal extraction capability was smaller with a thinner layer of PCM, and the higher return water temperature resulted in a larger amount of room heat gains. And the corresponding small drop in room heat losses with the increase in PCM layer thickness was a result of the thermal release from the PCM layer during the sunset hours. For office use, the room heat gains increased with the increase in PCM layer thickness because the thermal energy stored in the previous day was released to the water layer gradually.

Table 4.6 Comparison of room heat exchanges with different PCM layer thickness

(kJ)

	Thickness of PCM layer/mm	12	10	5
Residence	Typical summer-heat gain	29114.9	29115.4	29120.4
	Typical winter-heat gain	14151.0	14152.1	14153.8
	Typical winter-heat loss	8064.7	8065.3	8066.3
Office	Typical summer-heat gain	29075.1	29075.5	29073.0
	Typical winter-heat gain	13818.0	13817.0	13814.8
	Typical winter-heat loss	7824.3	7824.3	7823.7

4.4 Summary

To fulfill the off-work hot water demand of residential building and to improve the hot water supply capacity for office use, PCM is incorporated in the heat exchanger of the liquid flow window system. Thermal performance evaluation was completed numerically with validated method. This has been shown effective in peak water heat gain shifting and heat transfer enhancement as well. A summary of the findings is given below:

(1) For both applications in residential and office buildings, the addition of PCM to the annular space between the hot and cold streams (Case 1) is found to result in a decrease in water heat gains and an increase in room heat gains. On the other hand, the room heat losses of Case 1 are found much smaller than those of Case 2 and Case 3 in the winter season. This larger amount of heat loss of Case 2 and Case 3 through the window system is advantageous for office use in subtropical climate like the situation of

China by offsetting the internal cooling loads from occupant, lighting and equipment.

(2) With the PCM added to the outermost layer (Case 2), the water heat gains can be increased effectively as compared to the double-pipe heat exchanger of Case 3. The increase in water heat gains of Case 2 as compared to Case 3 is found larger for office use as compared to residential use during both typical summer and winter weeks. In typical summer week, the increases are 3.81% and 1.63% respectively for office and residential use. And the increase in typical winter week is less than 1%. It is indicated that the heat transfer enhancement effect of PCM application is more evident at higher cold feed water velocity.

(3) By incorporating PCM into the traditional double pipe heat exchanger, more hot water can be supplied during off work hours in residential building. Such load shifting can be more effective in the absence of domestic hot feed water supply during the daytime. But the indoor heat gains will be increased. It should be assessed properly case by case according to the actual hot water demand of the users.

(4) The proper amount of PCM to be used is affected by the ambient temperature and solar conditions. An increase of PCM layer thickness from 5mm to 10mm contributes to a larger increase in water heat gains, while the further increase of PCM layer thickness to 12mm has almost no significant contribution to the overall performance improvement in typical summer week. But a PCM layer of 5mm is enough for both residential and office use in the typical winter week. And the variation in the thickness of PCM layer had negligible impact on the thermal transmission through the window system.

5 Liquid Flow Window Performance under Different Climate Applications

Energy saving performance of water flow window in cooling application was found promising, so it is applicable for cooling-dominated climates. However, considering the diversity of weather conditions from area to area, studies under a full range of climate conditions are necessary. It is a well-known fact that in cold winter, water-freezing and ice formation may damage the system equipment and pipework. Various freeze prevention technologies are in use for solar energy applications. For example, transparent insulation material(TIM) is commonly used in flat-plate solar collectors. But the visual light transmission is reduced considerably under such condition, and extra thickness and weight are put on the system. Enhanced thermal insulation is another popular anti-freeze method. One example related closely to LFW application is the inclusion of a thermally insulated glazing system between two liquid chambers——known as a 'fluidized glass façade' in a recent European Commission project. With the liquid filled in outer chamber drained in winter, solar energy can be transmitted into the room space directly through the inner glazing. At the same time, warm water at a controlled temperature is circulated in the inner liquid chamber, so that the glazing itself can be used as a room heating device to maintain a good thermal comfort in indoor environment. This type of auxiliary heating is an 'active' (because of the power demand) but effective means in water freezing prevention. Alternatively, the use of anti-freeze additive like glycol is a popular 'passive' measure that can be used to lower the ice point of the circulating water.

Two types of glycols are popular in freeze prevention. They are: (1) the ethylene glycol (EG) that has better thermo-physical properties, and (2) the propylene glycol (PG) which is less toxic. Norton et al. studied the impact of PG concentrations on the freeze protection effect. It was concluded that 25% PG could be desirable for the city of London (with temperate oceanic climate), and a higher ratio was needed for places with winter temperature below $-10℃$. These were at the expense of drop in thermal conductivity and thermal efficiency.

As far as LFW is concerned, the use of TIM structure is undesirable because of the natural light blockage. And the auxiliary heating will cause the extra power demand. Glycol solution is then adopted as the practical passive means. Comparing with

EG, the safer PG is considered for window application.

5.1 Research Methodology

The study was completed numerically with a self-developed FORTRAN program. Simulations were completed to evaluate the application potential of liquid flow window under different climate conditions. In this study, seven climate regions from Regions I to VII in China were considered. One typical city was chosen in each of the climate zones to study and compare the thermal performance of the liquid-flow window application. Climatic characteristics, together with the proper PG concentration in each of the cities are given in Table 5.1.

Table 5.1 Representing cities in seven climate regions of China

Region	Description	Representing City	Location	PG conc.
I	Extremely cold $T_{mean} < -10°C$ in Jan.; $T_{mean} < 25°C$ in Jul.	Shenyang(SY)	42.41°N 123.29°E	40%
II	Cold weather $-10°C < T_{mean} < 0°C$ in Jan.; $18°C < T_{mean} < 28°C$ in Jul.	Beijing(BJ)	39.87°N, 116.43°E	15%
III	Hot summer and cold winter $0°C < T_{mean} < 10°C$ in Jan.; $25°C < T_{mean} < 30°C$ in Jul.	Shanghai(SH)	31.18°N, 121.48°E	0%
IV	Hot summer and warm winter $T_{mean} > 10°C$ in Jan.; $25°C < T_{mean} < 29°C$ in Jul.	Guangzhou(GZ)	23.02°N, 113.03°E	0%
V	Warm climate $0°C < T_{mean} < 13°C$ in Jan.; $18°C < T_{mean} < 25°C$ in Jul.	Kunming(KM)	25.37°N, 102.83°E	0%
VI	Cold and extremely cold $-22°C < T_{mean} < 0°C$ in Jan.; $T_{mean} < 18°C$ in Jul.	Lhasa(LS)	29.97°N, 91.11°E	15%
VII	Cold and extremely cold $-20°C < T_{mean} < -5°C$ in Jan.; $T_{mean} > 18°C$ in Jul.	Tikanlik(TK)	40.63°N, 87.70°E	40%

5.1.1 Model Validation for Anti-freezing Liquid Flow Window

The heat transfer mechanism of an anti-freezing liquid flow window is the same as that of water flow window. So, the same mathematical models for the water flow window were used in the validation test of the case with anti-freezing liquid as working fluid. Measured data including horizontal solar radiation, ambient and indoor air temperatures, as well as inlet temperature and flow rate of cold feed water acquired in the experiment were used in the validation test, by taking their hourly averaged values as the inputs.

In this study, the flowing water layer was replaced by anti-freezing liquid the same as that used in the experiment. The thermo-physical properties of the anti-freezing liquid with 40% propylene glycol are listed in Table 5.2. The variation in freezing and boiling points of aqueous solution of propylene glycol with concentrations is given in Table 5.3. It can be seen that the thermo-physical properties of anti-freezing liquid are different from that of water, so new relationships were built for the thermo-physical properties against the fluid temperature in FORTRAN program.

Table 5.2 Variation of thermo-physical properties of anti-freezing liquid with 40% propylene glycol

Temperature/℃	-15	0	10	20	30	40	50	60
Density/kg·m^{-3}	1050.43	1045.12	1040.94	1036.24	1031.03	1025.3	1019.06	1012.3
Specific heat/kJ·(kg·K)$^{-1}$	3.586	3.636	3.669	3.702	3.735	3.768	3.801	3.834
Thermal Conductivity/W·(m·K)$^{-1}$	0.369	0.385	0.394	0.402	0.409	0.415	0.42	0.425
Dynamic viscosity/m·Pa·s	33.22	12.37	7.22	4.57	3.09	2.21	1.66	1.3

Table 5.3 Freezing and boiling points of aqueous solution of propylene glycol

Percentage of Propylene Glycol		Freezing Point/℃	Boiling Point/℃
By Mass	By Volume		
0	0	0	100
5	4.8	-1.6	100
10	9.6	-3.3	100
15	14.5	-5.1	100
20	19.4	-7.1	100.6
25	24.4	-9.6	101.1
30	29.4	-12.7	102.2
35	34.4	-16.4	102.8
40	39.6	-21.2	103.9
45	44.7	-26.7	104.4
50	49.9	-33.5	105.6
55	55	-41.6	106.1
60	60	-51.1	107.2

The thermal absorption coefficient of the aqueous solution of propylene glycol was smaller than that of water slightly. For example, the thermal absorption coefficients of water and propylene glycol at a thickness of 10mm were 0.1357 and 0.09062, respectively. And thus, for the 10mm anti-freezing liquid with 40% propylene glycol, the absorption coefficient was taken as 0.118, corresponding to 88% of the absorption coefficient of water layer with the same thickness. The thermal absorption coefficient of the 10mm water layer is close to that of 2mm clear glazing, and consequently, the absorption coefficient of the 20mm water layer was taken the same as that of 4mm clear glazing. In this study, the thermal absorption coefficient of 0.165 was used for the anti-freezing liquid with a thickness of 20mm. It was taken as 88% of the solar absorption coefficient of 20mm water layer.

Uncertainty analysis was conducted and the error bands were determined using the same approach adopted in the validation test of the water flow window. The comparisons of simulation and experimental results are shown in Fig. 5.1 to Fig. 5.3. Temperatures at

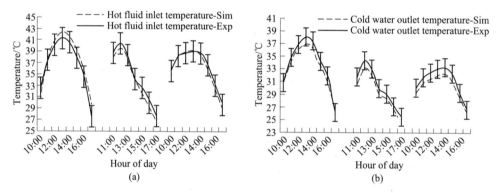

Fig. 5.1 Comparison of fluid temperatures in experiment and FORTRAN computation from 10am to 6pm for the anti-freezing liquid flow window
(a) Hot fluid inlet; (b) Cold water outlet

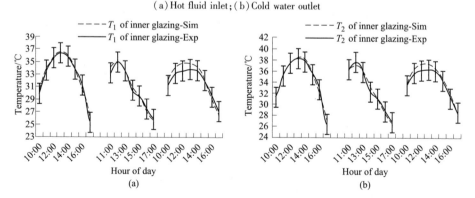

Fig. 5.2 Comparison of inner glazing surface temperatures in experiment and FORTRAN computation from 10am to 6pm for the anti-freezing liquid flow window
(a) T_1; (b) T_2

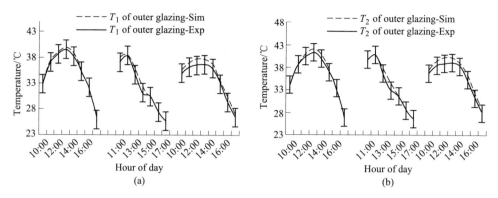

Fig. 5.3 Comparison of outer glazing surface temperatures in experiment and FORTRAN computation from 10am to 6pm for the anti-freezing liquid flow window
(a) T_1; (b) T_2

10am on Day 5 were excluded since the water supply was stopped before that time. The experimental and simulation results were found agreed well with each other.

5.1.2 Numerical Studies in Different Climates

With the validated model, more simulation runs were then conducted to evaluate the energy performance of LFW in each of the seven representing cities. Their yearly average solar intensities and the minimum ambient temperatures are listed in Table 5.4.

Table 5.4 Yearly average solar intensities and the minimum ambient temperatures

City	Shenyang (SY)	Beijing (BJ)	Shanghai (SH)	Guangzhou (GZ)	Kunming (KM)	Lhasa (LS)	Tikanlik (TK)
Solar intensity/kW·m^{-2}	1314	1402	1270	1139	1533	2032	1647
Ambient temperature/℃	-23.4	-14.2	-4.5	4.7	-2.1	-13.3	-20.1

It can be seen that, the lowest ambient temperatures in Shenyang and Tikanlik are −23.4℃ and −20.1℃ respectively. These are −13.3℃ and −14.2℃ in Lhasa and Beijing. Kunming and Shanghai are warmer with the lowest ambient temperatures of −2.1℃ and −4.5℃. But for most of the time, the ambient temperatures in Kunming and Shanghai are well above zero.

The required antifreeze liquid concentration is generally related to the local ambient temperature and determined by the lowest operating temperature. Guangzhou has no risk of freezing and thus water can be used as the working fluid. For the remaining six cities, liquid layer temperatures in the cavity are in practice well above the ambient temperature according to the simulation results. For example, it was found well above 5℃ in Shanghai and Kunming, above −5℃ in Beijing and Lhasa, and well above −15℃ in Shenyang and Tikanlik. Then there was no need to use glycol for Shanghai and Kunming, 15% concentration was appropriate for Beijing and Lhasa in order to reduce the freeze temperature to −5.1℃. Also 35% was good for Shenyang and Tikanlik as given in Table 5.1.

In this study, the effect of anti-freezing liquid on the system performance was first evaluated. To carry out such year-round performance evaluation with either water or anti-freezer as the working fluid, a city with no freezing risk had to be chosen. Thus Guangzhou in Region IV was selected as the illustrating example. The selected PG concentrations are from 15% to 35% by volume, with 5% increment. The varied absorption coefficients of the anti-freezing liquid layer were determined and given in Table 5.5.

Table 5.5 Thermal absorption coefficients of anti-freeze with different concentrations

Concentration/%	15	20	25	30	35
Absorption coefficient	0.178	0.175	0.171	0.168	0.165

The varied thermal physical properties, including density, viscosity and thermal conductivity of anti-freeze liquid with different concentrations are given in Table 5.6. The prediction was then followed by the year-round simulations in all the other climate regions. The TMY (typical meteorological year) weather data sets available from the VS-DOE website were adopted in this exercise.

Table 5.6 Varied thermal physical properties against liquid concentration

	Density/kg · m^{-3}									
Vol.%	−15	−10	−5	0	10	20	30	40	50	60
15%	—	—	1021	1020	1017	1014	1010	1005	1000	995
20%	—	1029	1027	1026	1023	1019	1015	1010	1005	999
25%	—	1034	1033	1031	1028	1024	1019	1014	1009	1003
30%	1041	1039	1038	1036	1033	1028	1024	1018	1013	1006
35%	1046	1044	1042	1041	1037	1032	1027	1022	1016	1009
	Conductivity/W · (m · K)$^{-1}$									
Vol.%	−15	−10	−5	0	10	20	30	40	50	60
15%	—	—	0.476	0.483	0.498	0.512	0.525	0.536	0.546	0.554
20%	—	0.442	0.449	0.456	0.470	0.483	0.494	0.505	0.514	0.521
25%	—	0.420	0.426	0.433	0.446	0.457	0.468	0.478	0.486	0.492
30%	0.391	0.397	0.403	0.409	0.421	0.431	0.441	0.450	0.457	0.463
35%	0.371	0.377	0.382	0.388	0.399	0.408	0.417	0.425	0.431	0.437

Continued Table 5.6

	Dynamic viscosity×10⁶/kg·(m·s)⁻¹								
Vol. %	-15	-10	-5	0	10	20	30	40	50	60
15%	—	—	4.055	3.365	2.34	1.72	1.315	1.035	0.84	0.7
20%	—	5.91	4.98	4.05	2.79	2.02	1.52	1.18	0.95	0.78
25%	—	8.875	7.025	5.560	3.655	2.540	1.855	1.405	1.105	0.895
30%	14.16	11.84	9.07	7.07	4.52	3.06	2.19	1.63	1.26	1.01
35%	23.615	17.475	12.85	9.685	5.865	3.825	2.655	1.935	1.47	1.16

5.2 Thermal Performance Analysis in Various Climate Applications

5.2.1 The Effect of Liquid Concentrations on thermal Performance

The comparison of the monthly variation in (1) the cold water heat gain at the heat exchanger, and (2) the thermal releases inward and outward from the window glazing surfaces, are shown in Fig. 5.4. The monthly average solar radiations are included as well. Both water heat gain and thermal release in the winter season are higher, the larger water heat gain is caused by the strong solar radiation and the low temperature of feed water in the winter season, while the larger amount of thermal release is mainly caused by the convection heat loss to the outdoor environment with low ambient temperature.

The addition of 15%, 20%, 25%, 30% and 35% of PG to the water layer brings about 12.66%, 17.62%, 22.62%, 26.74% and 30.60% reductions in annual water heat gains, respectively. Hence the larger the concentration, the less the water heat gain, and the lower the thermal efficiency. The result is in line with the previous findings of Norton et al.. Four typical months in different seasons were chosen to analyze the variation of water heat gains with liquid concentrations as indicated in Fig. 5.4(b). The water heat gain reduced linearly with the increase of glycol liquid concentration from 15% to 35% throughout the year. And the larger the useful water heat gain, the larger the discrepancy in water heat gain was observed with the increase of glycol concentration, as indicated by the different slopes.

The thermal release from the window system to the indoor and outdoor environments is increased along with the decrease in thermal extraction of the cold feed water. The difference is about 1.66% at 15% concentration, and with the further increase of liquid concentration, the difference is enlarged and reached 2.24%, 2.87%, 3.32% and 3.78% at concentrations from 20% to 35% with 5% interval.

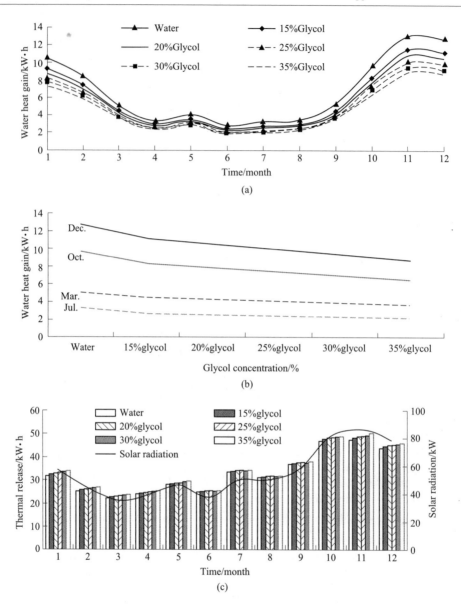

Fig. 5.4 Thermal performance comparison with various working fluids
(a) Monthly water heat gain with different concentrations;
(b) Water heat gain variation with glycol concentration during typical months;
(c) Thermal release inwards and outwards

5.2.2 Thermal Performance Comparison under Different Climates

Summary of yearly solar radiations incident on the 0.96m² glazing surface and the yearly average thermal efficiency in the seven typical cities are given in Table 5.7. High thermal efficiencies are achieved in almost all the seven cities. The yearly average thermal efficiency is as large as 14.21% in Lhasa, which is located in Region Ⅵ, though 15% of PG is mixed with water because of its strongest solar radiation. The comparative thermal efficiency of 15.62% in the city of Kunming located in Region V is also high, taking the fact that the solar radiation is strong and the air temperature is high during the daytime. The thermal efficiencies of Shanghai and Beijing in Regions Ⅲ and Ⅱ are found 14.7% and 12.02% respectively, while anti-freeze liquid with 15% of PG is used in the case study of Beijing. And this is caused by the higher radiation level in Beijing as compared to Shanghai. Thermal efficiency in extremely cold climate is not as high as that in cold weather. The slightly lower values of Shenyang in Region I and Tikanlik in Region Ⅶ as compared to the other cases are caused by the use of aqueous solution with larger concentration of 40% and their extremely low temperature in the winter season. In Guangzhou, which has similar climate to Hong Kong, thermal efficiency of 12.24% is achieved. This is 13.86% lower than that of Lhasa. It is indicated that once the freezing problem is solved, promising energy saving can be realized even in areas with cold climates but strong solar radiation.

Table 5.7 Yearly solar radiation and average thermal efficiency in different climates

City	Shenyang (SY)	Beijng (BJ)	Shanghai (SH)	Guangzhou (GZ)	Kunming (KM)	Lhasa (LS)	Tikanlik (TK)
Solar radiation/kW·h	937	1050	757	667	968	1315	1165
Thermal efficiency/%	8.58	12.02	14.70	12.24	15.62	14.21	8.74

The energy performance is evaluated in an integrated manner considering the water heat gain and the thermal transmission through the window. The seasonal results are given in Table 5.8.

Table 5.8 Comparison of seasonal thermal transmission and useful water heat gain (kW·h)

City		Shenyang (SY)	Beijng (BJ)	Shanghai (SH)	Guangzhou (GZ)	Kunming (KM)	Lhasa (LS)	Tikanlik (TK)
Water heat gain	Spring	23	32	25	12	26	32	20
	Summer	13	14	12	10	26	25	10
	Autumn	31	43	37	28	44	65	40
	Winter	13	36	37	32	55	65	33
	Sum	80	126	111	82	151	187	102

Continued Table 5.8

City		Shenyang (SY)	Beijing (BJ)	Shanghai (SH)	Guangzhou (GZ)	Kunming (KM)	Lhasa (LS)	Tikanlik (TK)
Room heat gain	Spring	19	34	20	42	35	10	40
	Summer	46	59	49	62	23	6	58
	Autumn	31	41	38	64	32	50	62
	Winter	6	25	15	31	51	60	36
	Sum	102	160	123	200	141	128	196
Room heat loss	Spring	37	21	25	7	19	46	30
	Summer	4	2	2	0	14	34	9
	Autumn	47	34	16	4	33	42	45
	Winter	111	72	51	21	36	57	86
	Sum	199	130	94	32	103	179	171

The useful water heat gain is affected by the weather condition and the indoor temperature. Thermal extraction capacity of the liquid layer is influenced by the solar radiation condition and the ambient and room temperatures. Under the stronger solar radiation and higher ambient temperature, both the direct and indirect thermal absorptions of the liquid layer increased. Then the flow viscosity can be smaller and the liquid circulation is improved. This will enhance the heat transfer at the heat exchanger from one aspect, and will reduce the friction loss along the flow path from the other.

The maximum year-round water heat gain is observed in Lhasa with the strongest solar radiation, followed by Kunming. The variation in water heat gain is in line with that of thermal efficiency. However, the water heat gain in Shanghai is lower than that of Beijing, though its thermal efficiency is higher, because the solar radiation level in Beijing is much higher, as indicated in Table 5.1. And the lowest water heat gain is observed in Shenyang because of its cold weather and weak solar radiation as explained above.

For most of the cities, the higher water heat gains are achieved during the autumn or winter season because of the strong solar radiation and the low cold supply water temperature. As mentioned above, the hot stream has higher temperature and flow velocity at the inlet of the heat exchanger under stronger solar radiation. And when the cold supply water is of low temperature, the temperature difference between the hot and cold streams is large. Then the greater temperature difference and flow velocity contribute to the improvement of the heat transfer effectiveness at the heat exchanger. The small water heat gain in the winter season of Shenyang is related to its extremely cold winter and the weak solar radiation level. For those cities like Beijing and Lhasa, the large amount of useful water heat gains during the winter season should also be attributed to the effective anti-freeze.

Thermal transmission to the indoor environment includes two parts: the direct solar transmission and the heat transfer from the glazing system. Thus, it is affected by both the solar condition and the ambient temperature. Besides, thermal release from the indoor environment to the outdoor environment through the glazing system takes place when the ambient is of very low temperature.

For most of the cities including Shenyang, Beijing, Shanghai and Guangzhou, the larger thermal transmissions to the indoor environment are in the summer season because of the high ambient temperatures during this period. While for Kunming and Lhasa with stronger solar radiations in the winter season, the maximum winter room heat gains are achieved. The high room heat gain during the autumn of Tikanlik is also caused by the strongest solar radiation during this period, and consequently the larger amount of direct solar transmission.

The yearly room heat gains in five out of the six cities (except for Tikanlik) are obviously lower than the cooling dominated Guangzhou. It is caused by their low ambient temperatures in the winter season. And the comparative room heat gains in Tikanlik and Guangzhou is caused by the large amount of direct solar transmission and the high ambient temperature during the summer and autumn periods in the desert area. Similarly, the low room heat gain in the summer season of Lhasa is mainly caused by its low radiation level during this period and the low ambient temperature of about 15.8℃ on average.

Heating is required in the winter season because of the room heat loss to the outdoor environment. It is negligibly small in the cooling dominated Guangzhou. The maximum room heat loss in Shenyang is caused by its extremely low ambient temperature in the winter season. In the summer season, the thermal release from the glazing system in Lhasa is even greater than that of Shenyang because of its low ambient temperatures in the night time.

The air-conditioning system consumes less electricity with a reduction in heat flow through the window system. The water heating system also consumes less when the useful water heat gain is higher. In this study, the corresponding electricity consumptions and savings led by a single piece of liquid flow window were estimated by taking the coefficient of performance (COP) of air-conditioning system as 3.5 for cooling and 4.5 for heating, and the efficiency of an electrical water heating system as 0.99. The year-round equivalent energy saving was then calculated by subtracting the electricity consumption for offsetting thermal transmissions through the window from the electricity saving of water heating system. The results are given in Table 5.9.

Table 5.9 Electricity consumption and saving caused by liquid flow window in different climate regions (kW)

City	Shenyang (SY)	Beijng (BJ)	Shanghai (SH)	Guangzhou (GZ)	Kunming (KM)	Lhasa (LS)	Tikanlik (TK)
Saving	81	128	112	82	153	189	103
Consumption	73	75	56	64	63	76	94
Equivalent saving	8	53	56	18	90	113	9

Saving in Tikanlik (Region Ⅶ) is 9kW, and that of Shenyang (Region Ⅰ) is 8kW. The consumption and saving are well balanced. Electricity savings in cities of Lhasa (Region Ⅴ) and Kunming (Region Ⅵ) are large with values of 113kW and 90kW respectively because of their large amount of water heat gains. Energy savings of 56kW and 53kW are achieved in Shanghai (Region Ⅲ) and Beijing (Region Ⅱ). Cities with no freezing risk are proven benefit a lot from this system, and promising energy savings are also achieved in those cities with strong solar radiation including Lhasa and Beijing. The better energy performance in Shanghai as compared to Shenyang is owing to its higher ambient temperature, despite that the solar radiation in Shanghai is weaker. Thus the energy performance of the system is promising in regions Ⅱ to Ⅵ except for the extremely cold regions of Ⅰ and Ⅶ. And the use of glycol is meaningful to maintain useful hot water generation in the winter seasons of regions Ⅱ and Ⅵ with cold winter.

5.2.3 Thermal Performance Improvement in Cold and Extremely Cold Climates

The use of glycol during the warm season resulted in a decrease in useful water heat gain, as indicated in the simulation results of Guangzhou. And part of the water heat gains in cities with cold climates may come from the indoor environment in the winter season when the ambient is of low temperature. It will also result in an increase in space heating load, especially for those cities with large heating demands.

Bearing in mind these two factors, the system performance in Shenyang, Beijing, Lhasa, and Tikanlik (i.e. cities of regions Ⅰ, Ⅱ, Ⅵ, and Ⅶ) with freeze prevention requirements were further studied. Firstly instead of anti-freeze, water was used as working fluid from Mar. to Oct. Secondly in the remaining months with freezing risk, two approaches given below were considered:

(1) Case 1: anti-freeze liquid was used.
(2) Case 2: the working fluid was drained.

The corresponding yearly water heat gains and thermal transmissions were compared. The results are given in Fig. 5.5 and Fig. 5.6.

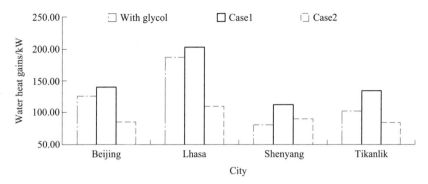

Fig. 5.5 Comparison of water heat gains of different cases

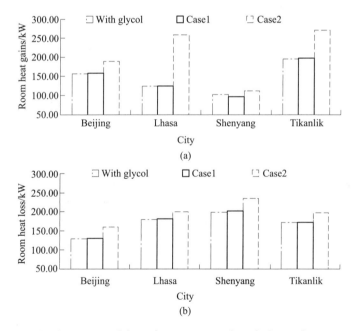

Fig. 5.6 Comparison of thermal transmissions through the window system
(a) Room heat gain; (b) Room heat loss

By using combined anti-freeze liquid and water as indicated in Case 1, the increases of water heat gains in Shenyang and Tikanlik are 39.88% and 31.41% respectively, and the corresponding increases in Beijing and Lhasa are 10.36% and 8.78% as compared to anti-freeze liquid flow window. Larger increases in Shenyang and Tikanlik located in Regions I and VII can be observed, because anti-freeze liquid with large concentration of 35% is needed for use in these two cities. And for cases with anti-freeze liquid of the same concentration, the larger differences in Shenyang and Beijing as compared to Tikanlik and Lhasa are resulted from their weaker solar radiation. Under the stronger solar conditions of Tikanlik and Lhasa, the impacts of the anti-freeze liquid concentration are less significant. From the point of useful thermal energy harvesting, the two cities of Shenyang and Tikanlik located in Regions I and VII benefit more through the replacement of anti-freeze liquid with water.

The overall water heat gains will be reduced with air-sealed window used in the winter season as indicated in Case 2. The reductions are significant for the cities of Lhasa, Tikanlic and Beijing, which means that the sacrifice of water heat gains is more evident for cities with strong solar radiation.

By replacing anti-freeze liquid with water during the warm period (Case 1), the thermal transmission is hardly affected. However, for Case 2 with water replaced by air in the winter season, both room heat gain and heat loss via the window system are increased because the thermal resistance is smaller without the radiation blocking. The increase in room heat gain of Shenyang located in Region I is not as great as the other three cities because the solar radiation in Shenyang is weak and the ambient temperature is low. However, the overall system performance of Case 2 with liquid drained in the winter season is not as good as the liquid filled window because of the missing of useful water heat gains.

The simulation results indicate that the thermal efficiency of the system is promising even in cold climates once the freezing problem is properly solved, especially for areas with cold climate but strong solar radiation. The anti-freeze is readily available on the market but it is of high cost, which should be taken into consideration in the practical use.

5.3 Summary

The model validation test for the self-developed computer program is successful, with good agreement between the experimental and simulation results. Then year-round simulation was conducted with the self-developed program for system performance evaluation. The impact of anti-freeze liquid concentration on the system thermal performance and the energy performance of liquid flow window for use in different climates are evaluated and the key findings and conclusions are given below:

(1) The use of anti-freeze liquid than water brings about a decrease in system water heat gain and thermal efficiency. But it increases the thermal releases inwards and outwards. The water heat gain reduced linearly with the increase of liquid concentration from 15% to 35% by 5% increment. The impacts depend also highly on the solar radiation level except the concentration of anti-freeze liquid in use. Under a stronger solar radiation, the impact of liquid concentration is less significant. With anti-freeze concentration of 35%, the difference in thermal efficiency is 31.41% under yearly solar radiation level of 1165kW, while it is 39.88% with the radiation level of 937kW. Similarly, with anti-freeze concentration of 15%, the differences of 8.78% and 10.36% are observed corresponding to solar radiation levels of 1315kW and 1050kW.

(2) Overall electricity consumptions of air-conditioning and hot water systems can be reduced effectively by installing liquid flow window. This is the fact in all climate regions of China. For those areas with freezing risk, promising energy savings of 113kW/piece/year and 53kW/piece/year in Lhasa and Beijing in regions Ⅵ and Ⅱ with cold climate can be realized by effective freezing prevention. The energy consumption and saving caused by liquid flow window are well balanced in extremely cold climates of Regions Ⅰ and Ⅶ.

(3) For use of the window system in Regions Ⅰ, Ⅱ, Ⅵ and Ⅶ with anti-freeze requirements, the replacement of anti-freeze liquid with water from Mar. to Oct. contributes to the improvement in system energy saving. But extra labor input is required for this operation and maintenance in real practice. A large amount of useful water heat gain is sacrificed if water is drained away in the winter season, and the thermal transmission is increased. Thus the overall energy performance of liquid filled system is more promising compared to the air sealed winter operation mode.

6 Life Cycle Assessment of Liquid Flow Window

For many innovative solar technologies, large amount of energy is often needed during the manufacturing process though their operations are considered to be cleaner than the traditional energy generation methods. Because of this, a growing number of research works have been on the comprehensive energy and environmental assessments of novel solar engineering systems, known as the Life Cycle Assessment (LCA). This is an integrated process to assess the economic, energy and environmental impacts associated with all stages of the product life span, by considering the activities from extraction of raw material, manufacturing, transportation, installation and commissioning, operation and maintenance, up to the final disposal after use.

More often, researchers cared about the economic impact of the new technologies, like in solar thermal and electricity systems. The cost benefit of Solar Water Heating System (SWHS) was studied by Diakoulaki et al. considering the energy saving, environment impact, and the corresponding working opportunities in Greece. Economic performance of solar water heating system for use in residential buildings in U. S. was also studied; it was concluded that the energy saving can be as high as 50%~85%, which varied according to different regions, and the corresponding annual savings for a typical home can be $100 to $300. A research in Singapore studied the cost effect of a solar water heating system and suggested that there existed an optimum scale for the installation of solar water heater. Life cycle saving and life cycle cost methods were adopted in their study and the cash flow discount was considered.

The energy and economic performance evaluation of many solar thermal systems have also been widely reported. To quote some examples, Mateus et al. performed the energy and cost analyses of an integrated solar system for space heating and cooling, under the different climate conditions of Berlin, Lisbon and Rome; the payback time was found location dependent. The energy payback time of a typical solar domestic hot water system was calculated in the study of Streicher et al. ; a payback period less than 2 years was achieved. A research in Switzerland studied the energy and environmental impacts of a solar thermal system that provided 100% space heating and hot water services. Comparing to the traditional systems, the purchase of primary energy was reduced by 84% ~ 93% and the reduction in green-house-gas emission was 59% ~ 97%. A reduction of carbon emission by 70% was reported in another study of solar thermo-syphon domestic water heaters. Similar integrated analysis was completed by Kalogirou from the economic, energy and environment prospectives.

Solar technologies carry impacts on the environment during every step of raw material extraction, transportation, manufacturing, on top of the normally considered supply and installation, as well as operation and maintenance processes. Facing the uncertainty during each step, sensitivity analysis was commonly adopted in the LCA exercise. For instance, uncertainties in the processes of production, operation and disposal of an integrated collector storage system were considered in the works of Battisti and Corrado; the sensitivity analysis was helpful to the eco-design of solar collectors. Ardente et al., conducted an integrated impacts evaluation of the solar thermal system from material extraction to final disposal. Accordingly, the uncertainties in data quality, means of transportation, maintenance and disposal were taken into consideration; the eco-profile from energy and material production to transportation and disposal was traced.

With more components involved in the advanced liquid flow window system, more fuel is consumed during the manufacturing process. Nevertheless, so far, the related studies on the comprehensive analysis on liquid flow window are very few. Thus, the objective of this study is to conduct an integrated life cycle assessment of the mentioned liquid flow window system from the economic, energy and environmental aspects.

There are totally four stages in LCA, including goal and scope definition, inventory analysis, impact assessment and improvement assessment. The goal and scope should be defined clearly in the first place in LCA study. Then the Life Cycle Inventory Analysis can be carried out to calculate the energy input and the greenhouse gas emission of a product from cradle to grave. In the last place, the life cycle impact can be evaluated based on its energy saving and emission reduction potentials. Life Cycle Interpretation is a systematic process to assess and summarize the results from other stages. With reference to the above, the boundary of the LCA analysis is first defined then followed by the life cycle economic, energy and environmental impacts assessment. Two different climate conditions, i. e. Hong Kong (Cooling dominated) and Beijing (Cooling in summer and heating in winter) are used for illustration. The contributions or disturbances of different material flows to the environment are analyzed and this is good for product and process improvement.

6.1 The Method of Analysis

Fig. 6.1 defines the boundary of this life cycle assessment study, with the impacts covering the economic, energy and carbon emission aspects. The Cost Pay-Back Time (CPBT), Energy Pay-Back Time (EPBT) and Green-House-Gas Pay-Back Time (GPBT) were determined for performance evaluation.

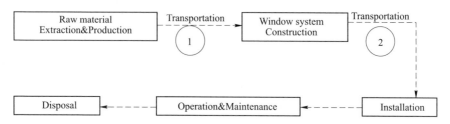

Fig. 6.1 Boundary of LCA of liquid flow window

CPBT was calculated by the estimation of the sum of the investment for construction of the system divided by annual saving of money for buying electricity less the incurred operation and maintenance expenses. EPBT was calculated by the life-long energy input divided by the annual energy saving because of the use of this system. In the same way, GPBT was the ratio of green-house gas emission during the life-long operation of the system to the yearly emission reduction. The emission reduction was calculated based on annual electricity saving, it was the value of carbon emission for generating the same amount of electricity with a local power plant. In this study, EPBT was calculated by

$$\text{EPBT} = \frac{E_{win} + E_{bos} + E_{tran} + E_{install} + E_{dis}}{E_t + E_{ac} - E_{om}} \tag{6-1}$$

Where E_{win} ——the overall embodied energy for the factory production of the main components of the window;

E_{bos} ——the embodied energy for the production of the balance of the system like accessory components;

E_{tran} ——the energy input for system equipment and materials transportation;

$E_{install}$ ——the energy consumption for on-site installation;

E_{dis} ——the energy used for decomposition and disposal;

E_t ——the annual energy saving of the water heating system in operation;

E_{ac} ——the annual energy saving of the air-conditioning system as a result of replacing the conventional window system with the liquid-flow window system;

E_{om} ——the annual energy consumed in association with the additional operation and maintenance work of the new window system as compared to the base-case system.

GPBT was calculated with formulas by

$$\text{GPBT} = \frac{G_{win} + G_{bos} + G_{tran} + G_{install} + G_{dis}}{G_t + G_{ac} - G_{om}} \qquad (6\text{-}2)$$

Where G_{win} ——the green-house gas emission corresponding to the embodied energy term E_{win};

G_{bos} ——the green-house gas emission caused by the balance of system term E_{bos};

G_{tran} ——the green-house gas emission during the transportation process;

$G_{install}$ ——the green-house gas emission during the on-site installation process;

G_{dis} ——the green-house gas emission during the process of decomposition and disposal;

G_t ——the reduction in green-house gas emission of the water heating system;

G_{ac} ——the reduction in green-house gas emission of the air-conditioning system;

G_{om} ——the annual green-house gas emission from the associated operation and maintenance activities.

The LCA of liquid flow window was completed based on its application in an integrated sport center under the weather conditions of Hong Kong and Beijing. An antifreezing liquid with 15% propylene glycol was used as working fluid for the application example in Beijing. Two window systems employing different glazing types were considered in this study. The base case was the 24mm commercial-grade IGU(with 6mm clear glazing+12mm air layer+6mm low-E glazing), and its energy performance for use in a sport center was compared with the double absorptive glazing liquid flow window. Side views of the two glazing systems are shown in Fig. 6. 2.

The purpose of the LCA study covered in this chapter is not to have a detailed and through analysis of the liquid flow window system for its general application in every part of the world including the environmental impact and risk. Instead, It serves to give an indication of the application potential based on a real building case with good demand and water services. The air-conditioning energy consumption in each functional space of the sport center is determined by the full input of building zones, construction materials and thermal properties, fresh air demands, internal loads and their operating schedules. The ESP-r platform was used for such whole-building energy use evaluation.

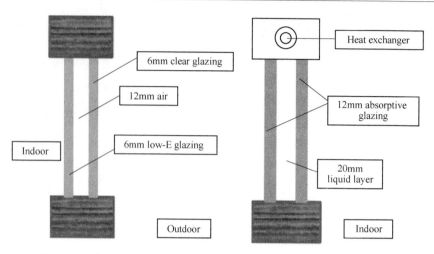

Fig. 6.2 Side view of IGU and double absorptive liquid flow window system

6.2 Life-long Investment, Energy Input and Carbon Emission of Liquid Flow Window

6.2.1 Investment, Energy Input and Carbon Emission for Window Materials

For the window prototype used in the experimental test, the window frame structures, distribution pipes and headers in the window cavity were made of stainless steel, and copper was used to construct the heat exchanger. The remaining small components for connection were primarily made of stainless steel. Thus, three primary kinds of materials were involved in the prototype construction. Materials consumptions by volume for the manufacture of a single piece of window system are given in Table 6.1.

Table 6.1 Materials consumptions for producing a single piece of liquid flow window

Component	Material	Volume/m^3
Glazing	Absorptive glazing	0.02304
U-channel	Stainless steel	0.000798
Window frame	Stainless steel	0.0029915
Distribution headers	Stainless steel	0.00026
Double-pipe heat exchanger	Copper	0.000155744
ϕ20mm circulation pipe	Stainless-steel	0.00015072
Pipe connections	Stainless steel	0.00028536
Insulation	Mineral wool	0.001884
Drain valve	Stainless steel	1 piece

According to the price provided by Alibaba, the price of IGU was about 160 CNY/m². This was ranged from 20 CNY/m² to 40 CNY/m² for absorptive glazing. Then the price of 30 CNY/m² was used for the absorptive glazing in the present study. The estimated cost of IGU with a surface area of 0.96m² was 154 CNY, and the cost for the two absorptive glazing panes was 58 CNY. On top of this, extra investments were required in liquid flow window system for pipes, frame structures and the heat exchanger, and these were taken as 400 CNY with stainless steel used as the frame material. Thus the extra investment for producing a single piece of liquid flow window as compared to the IGU was about 304 CNY. The working fluid used in this system also contributed to extra cost. The distilled water was cheap, but the price of anti-freezing liquid was much higher. And the extra investments of the working fluids were 15 CNY and 240 CNY respectively for Hong Kong and Beijing. Anti-freezing liquid was used in Beijing for freezing prevention. The corresponding extra investments in sum were 319 CNY and 544 CNY respectively.

The embodied energy and the corresponding carbon emission for materials used in this window system were calculated based on the database of eBALANCE developed by IKE company in China. Three databases including the China Life Cycle Database (CLCD), the Europe Life Cycle Database (ELCD), and the Ecoinvent database of Switzerland were integrated in this software. Energy consumption and equivalent carbon emission from cradle to gate of each material were provided.

The absorptive glazing used in the liquid flow window was a kind of flat glass, and the energy consumptions for the production of 1kg of flat glazing are listed in Table 6.2. The

Table 6.2 Energy input and carbon emission for production of 1kg of flat glazing

Energy input		Carbon emission			
Energy source consumed	Energy input /MJ	Emission product	Emission of unit product/kg	CO_2 equivalent factor	Total emission of CO_2/kg
Brown coal	0.4720	CO_2	0.743	1	0.743
Crude oil	1.9100	NO	0.000007240	298	0.002158
Hard coal	1.5600	Methane	0.00149	25	0.03725
Natural gas	5.6200	—	—	—	—
Peat	0.0087	—	—	—	—
Geothermic	0.0088	—	—	—	—
Hydro power	0.2130	—	—	—	—
Solar energy	0.0405	—	—	—	—
Waves	0.0000	—	—	—	—
Wind power	0.0456	—	—	—	—
Sum	9.88	—	—	—	0.78

energy inputs and carbon emissions were determined based on the ELCD3.0 database. For the production of flat glass, the resources consumed were mostly traditional fossil fuels, but renewable energy like wind, hydro and solar energy were also used. The total energy input was about 9.88MJ for producing 1kg of flat glazing. The corresponding equivalent emission of carbon was about 0.78kg CO_2-eq. For the emission of nitrogen oxides and methane, they could be taken as the equivalent emission of CO_2, and the corresponding transition factors were 298 and 25 respectively.

Energy consumption for the production of stainless steel and copper could be calculated in the similar way. The stainless steel used in the prototype was steel plate of 2mm thick, and the corresponding energy input and carbon emission is given in Table 6.3. For the production of steel plate, the energy sources were mainly fossil fuels like crude oil and hard coal. The total energy consumption for the production of 1kg of steel plate was about 9.68MJ, and the corresponding green-house gas emission was about 1.35kg CO_2 equivalently. This was greater than the carbon emission for the production of glazing as a result of more fossil fuel consumption.

Table 6.3 Energy input and carbon emission for production of 1kg of steel plate

Energy source	Embodied energy		Carbon emission		
	Energy input /MJ	Emission product	Emission of unit product/kg	CO_2 equivalent factor	Total emission of CO_2/kg
Crude oil	1.03	CO_2	0.9109	1	0.9109
Hard coal	6.19	NO	0.0014073	298	0.4193754
Natural gas	2.46	Methane	0.0007050	25	0.017625
Sum	9.68	—	—	—	1.35

For the copper material used in the heat exchanger, the data of energy input and carbon emission were available in the ELCD3.0 database and the corresponding energy consumption and carbon emission for the production of 1kg of copper is given in Table 6.4. Both traditional fossil fuel and renewable energy were used and the total energy input for producing 1kg of copper pipe was 18.3MJ. The corresponding carbon emission was about 0.98kg CO_2 equivalently. The energy consumption for the production of copper products was much greater than the other two kinds of materials. But the carbon emission was quite similar to the other two materials since the amount of fossil fuel consumption was close.

Table 6.4 Energy input and carbon emission for production of 1kg of copper pipe

Energy input		Carbon emission			
Energy source	Energy input/MJ	Emission product	Emission of unit product/kg	CO_2 equivalent factor	Total emission of CO_2/kg
Brown coal	1.71	CO_2	0.927	1	0.927
Crude oil	2.00	NO	0.000021500	298	0.006407
Hard coal	3.50	Methane	0.00188	25	0.047
Natural gas	5.17	—	—	—	—
Hydro power	0.69	—	—	—	—
Wind power	0.02	—	—	—	—
Uranium	5.14	—	—	—	—
Wood	0.11	—	—	—	—
Sum	18.34	—	—	—	0.980407

The difference in energy input for each material production was related to the energy source condition of different databases and the varied manufacturing approaches. Based on the above-mentioned results of energy inputs of different materials, the embodied energy for the materials in construction of liquid flow window system could be determined. The cumulative energy input and the corresponding carbon emission of materials used for the construction of a single piece of liquid flow window are given in Table 6.5 and Table 6.6.

Table 6.5 Cumulative energy input for materials used for construction of a single piece of liquid flow window with stainless steel frame

Materials/Components		Quantity consumed/kg	Energy intensity /$MJ \cdot kg^{-1}$	Cumulative energy/$kW \cdot h$
Glazing	Absorptive glass	57.95	9.88	159.02
Frame	Stainless-steel frame	30.05	9.68	80.80
Heat exchanger	Copper pipe	2.33	18.34	11.84
Connection pipe and accessories	Distribution headers	4.37	9.68	11.74
	Connection pipes	1.86	9.68	5.01
	Flange	1.59	9.68	4.29
	Valve	0.50	9.68	1.34
	Insulation (glass wool)	0.1884	31.7	5.97
Sum		98.84	—	280.02

Energy consumption of the glazing panes contributed to about 57% of the extra embodied energy. The frame structure contributed to a smaller portion of about 29% to the total energy input, and these two main parts together resulted in about 86% of the total energy consumption. The heat exchanger contributed to only about 4.23% of the energy consumption because of its relatively light weight though the energy input for the production of copper pipe was large. Distribution headers in the window cavity consumed the larger amount of energy as compared to the heat exchanger because of their overall weights. The cumulative embodied energy was 280.02kW · h for the construction of a single piece of liquid flow window.

Table 6.6 Cumulative carbon emission for materials used for construction of a single piece of liquid flow window with stainless steel frame

Materials/Components		Quantity consumed/kg	CO_2 emission rate /kg CO_2-eq · kg^{-1}	CO_2 emission /kg CO_2-eq
Glazing	Absorptive glass	57.95	0.78	45.34
Frame	Stainless-steel frame	30.05	1.35	40.51
Heat exchanger	Copper pipe	2.33	0.98	2.28
Connecting pipe and accessories	Header	4.37	1.35	5.88
	Connections	1.86	1.35	2.51
	Flange	1.59	1.35	2.15
	Valve	0.50	1.35	0.68
	Insulation(glass wool)	0.1884	2.80	8.80
Sum		—	—	108.15

Carbon emissions for materials production varied with different materials, and the emission for producing absorptive glazing was slightly smaller than the other materials. Thus the contribution of the glazing panes to carbon emission was about 42%, while the contribution of the frame was 37%, and that of the headers was also increased. The emission in association with the heat exchanger was reduced because of the lower emission rate during the copper production process. And the total equivalent emission of CO_2 for the construction of a liquid flow window was 108.15kg CO_2-eq.

The energy input and carbon emission for the construction of the liquid flow window varied with the material selection. They could be further reduced from two aspects, one way was the adoption of glazing material with smaller thickness and the other one was the replacement of the frame material because they were the main sources of energy consumption and carbon emission. Thickness of the glazing panes should be determined carefully because deformation or even breakage might happen if the glazing was not strong enough. However, alternative materials like wood could be a good choice for the frame structure because of its light weight, minor environmental impact and short recovery period. The cost for the wooden frame was also reduced to 180 CNY. The extra investments were then 99 CNY and 324 CNY respectively in Hong Kong and Beijing. While the energy input for producing 1kg of wood was about 10.8MJ, which had no big difference comparing with other materials, but the overall weight of a wooden frame could be substantially reduced because of its much smaller density as compared to a metallic frame. The corresponding carbon emission was also much lowered at 0.065kg for producing 1kg of wood. The absorptive glazing then became the main source of energy consumption and carbon emission in the case of wooden frame, the energy consumption and the corresponding carbon emission for the window materials could be reduced to 238.73MJ and 60.98kg CO_2-eq. The reductions in cumulative embodied energy and carbon emission were 15% and 44% respectively as compared to those with stainless steel frame.

For the application in Beijing, the anti-freezing liquid also led to extra energy consumption and carbon emission. They were estimated as 69kWh and 5.76kg CO_2-eq based on the consumption and emission rates of 13MJ/kg and 0.3kg CO_2-eq/kg. Thus, the embodied energy with stainless steel and wooden frames were 349.02kW·h and 307.73kW·h. Correspondingly, the carbon emissions were 113.91kg CO_2-eq and 66.74kg CO_2-eq respectively.

6.2.2 Cost, Energy Consumption and Carbon Emission from Construction to Disposal

As indicated in the boundary designation (shown in Fig. 6.1), cost, energy input and carbon emission during the processes of transportation, assemble or construction, installation, operation and maintenance as well as disposal were included. The negative effects during the installation process were found to be negligibly small, and so this was excluded from the analysis.

In the construction of liquid flow window, forming/extrusion technique was used for the production of separate components like U-channel, window frame, pipe connection and distribution headers. These were made of stainless steel. Stainless steel water pipe and absorptive glass materials were readily available in the market. Headers, connections and circulation pipes were made connected to one another by welding. Copper pipes used for the heat exchanger were also available in the market and welding was needed for its fabrication. An indicative diagram of the Life-Cycle inventory of the liquid flow window is shown in Fig. 6.3 for reference.

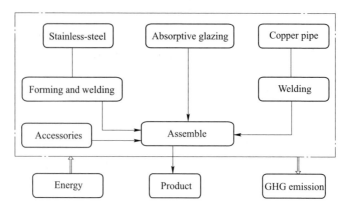

Fig. 6.3 Life-Cycle inventory diagram of liquid flow window

Traditionally, the major construction processes for window are sealing and assembly, while the construction of liquid flow window is more complicated because of the setting of pipe-work. The average equivalent manufacturing period for a single piece of liquid flow window is 2h with a single worker. Since the IGU is a commercialized product, the manufacturing cost is readily included in its price of 160 CNY/m². Thus, the extra cost for liquid flow window manufacturing comparing to IGU should be 62.5 CNY. It was calculated based on a daily (8) salary of 250 CNY/Person. Traditionally, the energy input for sealing was estimated at 1.7kW·h, and it was applicable for both water flow window and IGU. But the assembly of liquid flow window is much more complicated. The consumption for assembly was 13.8kW·h with wooden frame; and for steel frame, it was assumed the same as that of aluminum frame, i.e., at 7.2kW·h. It was taken as 1.5 times of the traditional assembly process considering the structure complexity of the liquid flow window. The corresponding carbon emission was determined based on the amount of end-use electricity. They were 4.8kg CO_2-eq and 2.9kg CO_2-eq with wooden and stainless-steel frames.

As for transportation, the corresponding energy consumption and carbon emission were determined based on mass and transportation distance. The energy consumption of an 8-ton trunk transportation in China was 2.05MJ/(t·km), and the corresponding carbon emission was 0.444kg CO_2-eq/(t·km). In this study, the prototype was manufactured in Dongguan of Guangdong province and the transportation distance from Dongguan to Hong Kong was around 180kg. For the case study of Beijing, it was assumed that the windows were manufactured in a nearby city, and the transportation distance to Beijing was similar to that from Dongguan to Hong Kong. The economic and environmental impacts during the disposal process were incurred mainly by the transportation process from the installation site, and the economic transportation distance should be less than 20km. Thus, a total transportation distance of 200 km should be taken into consideration for product delivery as well as disposal process. The transportation cost was determined by the mass of product goods and the transportation distance. This was at the price of 0.6 CNY/kg for the product delivery distance of 180 km.

Operation was not considered in this analysis since no pump or extra power would be required in the operating stage of the liquid flow window. Extra maintenance efforts would be required for leakage checking, heat transfer fluid supplement or replacement. With the assumption of two replacements, the cost, energy input and carbon emission were respectively 480 CNY, 139kW · h and 12kg CO_2-eq for anti-freezing liquid. In Hongkong, water is used as liquid medium. Since it is widely available, and the corresponding energy input and carbon emission was assumed to be negligible, thus excluded from the current study.

However, energy consumption and carbon emission also incurred during the construction process of the conventional low-E glazing. The thickness of the low-E glazing is half of the glazing panes used in the liquid flow window, and thus the corresponding energy input and carbon emission can be reduced by half. Its energy input and carbon emission are then 79.51kW · h and 22.67kg CO_2-eq. While the construction of liquid flow window is more complex and it is heavier with extra materials for pipework. The calculated cost, energy input and carbon emission for materials production, window construction, transportation (including disposal) and maintenance of liquid flow window and IGU are summarized in Table 6.7 and Table 6.8. Different types of frame materials were taken into consideration. The use of anti-freezer in Beijing causes the higher cost than in Hongkong and brings about extra investment, energy consumption and carbon emission during the maintenance process.

Table 6.7 Lifelong Cost, energy input and carbon emission for window materials in Hong Kong application

Input	Items		Liquid flow window	IGU	Difference
Cost/CNY	Material	Steel frame	473	154.00	319.00
		Wooden frame	253		99.00
	Construction		—	—	62.50
	Transportation (including disposal)	Steel frame	69	20.28	48.90
		Wooden frame	37		16.79
	Maintenance		30	0	30.00

Continued Table 6.7

Input	Items		Liquid flow window	IGU	Difference
Energy input /kW·h	Material	Steel frame	280.02	79.51	200.51
		Wooden frame	238.73		159.22
	Construction	Steel frame	7.2	4.80	2.40
		Wooden frame	13.8	9.20	4.60
	Transportation (including disposal)	Steel frame	41	11.88	28.64
		Wooden frame	22		9.84
Carbon emission /kgCO$_2$-eq	Material	Steel frame	108.15	22.67	85.48
		Wooden frame	60.98		38.31
	Construction	Steel frame	2.9	1.93	0.97
		Wooden frame	4.8	3.20	1.60
	Transportation (including disposal)	Steel frame	9	2.57	6.20
		Wooden frame	5		2.13

Table 6.8 Lifelong Cost, energy input and carbon emission for window materials in Beijing application

Input	Items		Liquid flow window	IGU	Difference
Cost/CNY	Material	Steel frame	698	154.00	544.00
		Wooden frame	478		324.00
	Construction		—	—	31.25
	Transportation (including disposal)	Steel frame	69	20.28	48.90
		Wooden frame	37		16.79
	Maintenance		480	0.00	480.00
Energy input /kW·h	Material	Steel frame	349.02	79.51	269.51
		Wooden frame	307.73		228.22
	Construction	Steel frame	7.2	4.80	2.40
		Wooden frame	13.8	9.20	4.60
	Transportation (including disposal)	Steel frame	41	11.88	28.64
		Wooden frame	22		9.84
	Maintenance		139	0.00	139.00
Carbon emission /kgCO$_2$-eq	Material	Steel frame	113.91	22.67	91.24
		Wooden frame	66.74		44.07
	Construction	Steel frame	2.9	1.93	0.97
		Wooden frame	4.8	3.20	1.60
	Transportation (including disposal)	Steel frame	9	2.57	6.20
		Wooden frame	5		2.13
	Maintenance		12	0.00	12.00

Based on this, the overall extra cost, energy input and carbon emission during the whole life span of a single piece of liquid flow window as comparing to IGU are summarized in Table 6.9.

Table 6.9 Extra life cycle cost, energy input and carbon emission for a single piece of liquid flow window

City	Investment/CNY		Embodied energy/MW·h		Equivalent carbon emission	
	Stainless steel frame	Wood frame	Stainless steel frame	Wood frame	Stainless steel frame	Wood frame
Hong Kong	460	208	232	174	93	42
Beijing	1135	883	440	382	110	60

6.3 Building Energy Simulation in ESP-r

Energy saving performance evaluation of liquid flow window in building application was completed with ESP-r. And a validation study was conducted in the first place by comparing to experimental results, followed by the year-round building energy simulation under climates of Hong Kong and Beijing.

6.3.1 Validation Study

To prove the readiness of ESP-r in energy performance evaluation of liquid flow window, a validation study was first conducted. It was used to model the thermal behavior of liquid flow window because liquid filled zones are supported in ESP-r. The validation study was conducted based on the different working fluids of water and anti-freezing liquid. A test room shown in Fig. 6.4 with the same dimension as the experiment chamber

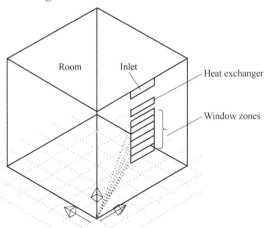

Fig. 6.4 Model of experiment chamber with liquid flow window developed in ESP-r

was built. Measured weather data from the experiment was imported into the ESP-r weather module. The solar absorption coefficients of the 20mm water layer and the antifreezing liquid used in the experiment were estimated at 0.187 and 0.165. The 1.2m high window was divided into six adjacent zones. Two more zones were added: one of them was used for modelling the heat transfer in the heat exchanger, and the other was taken as a dummy zone to store the inlet temperature of cold feed water.

The flow network module of node and component structure in ESP-r could be used for flow rate calculation. In this study, nodes were built for the six window zones and the heat exchanger zone. Friction loss along the flow path was modelled via the interconnected components. The structure of the flow network created in this study is shown in Fig. 6.5. In the window cavity, the friction loss factor and dimensions along the flow path of the six window zones were the same, and thus the same component of component 1 was used. Two other components were built for the calculation of friction loss from the top window zone to the heat exchanger and from the heat exchanger to the bottom window zone, corresponding to component 2 and component 3.

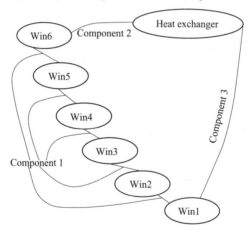

Fig. 6.5 Flow network of liquid flow window in ESP-r

By connecting nodes in the flow network module to the corresponding zones of window and the heat exchanger in the energy simulation module, communications between the two modules were thus set, temperature and velocity values could then be shared. The path of solar transmission was adjusted to make it reach the inner glazing through the liquid layer. Thus, a part of the solar thermal energy was made absorbed by the flowing liquid stream. The convective heat transfer coefficients at the glazing surfaces were assigned since the heat transfer coefficients were determined taking the working fluid as air by default in ESP-r. The temperature-dependent physical properties of the working fluids were input at the source code level.

The comparisons of glazing surface temperatures from simulation and experimental test introduced in Chapter 2 are shown in Fig. 6.6 and Fig. 6.7. The good agreement between them confirms that ESP-r is able to predict the thermal performance of liquid flow widow with enough accuracy. Integrated building energy simulation runs were then conducted accordingly.

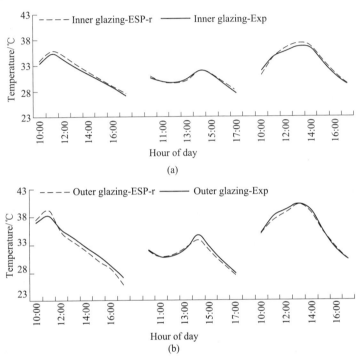

Fig. 6.6 Comparison of glazing temperatures in experiment and ESP-r simulation for the water flow window
(a) Inner glazing; (b) Outer glazing

6.3.2 Building Energy Simulation

The case study was completed based on the application of water flow window in an integrated sport center with nine function rooms, the distribution of the function rooms are shown in Fig. 6.8. Windows were provided on the vertical walls and also at the tilted rooftop of each room, except for the plant room and the changing room. These two rooms were thus excluded from the building energy simulation. The total number of windows installed was 400 pieces, corresponding to a total glazing area of 384m^2 based on the window prototype of 0.96m^2 each.

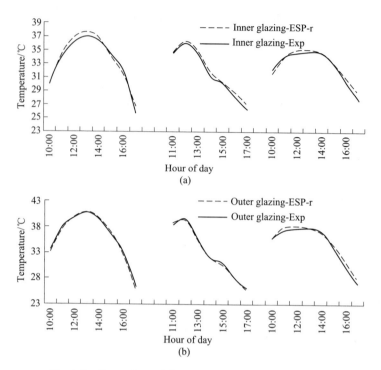

Fig. 6.7 Comparison of glazing temperatures in experiment and ESP-r simulation for the anti-freezing liquid flow window
(a) Inner glazing; (b) Outer glazing

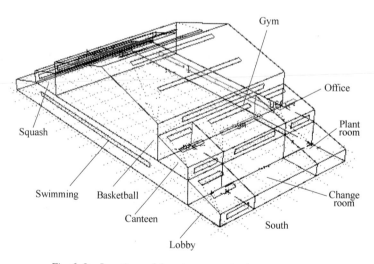

Fig. 6.8 Locations of function rooms in the sport center

In the building energy simulation with ESP-r, representative computer models of the sport center were developed in the first place. The group of small window units at the same row was replaced by one piece of large window with the equivalent glazing area.

For the following six function rooms namely main lobby, canteen, gym room, office, indoor swimming pool and badminton game hall, two pieces of large windows were located on the roof top surfaces and also the vertical walls. The number of large windows for the basketball game hall was six; four of them were located on the roof top and the other two were installed at the south-facing and north-facing vertical walls. Each of these equivalent large windows was divided into four zones. A heat exchanger zone was then incorporated for determining the heat release from the hot fluid. There were totally 9 building zones and 90 window-related zones in the model of the sport center. The whole model was split into two parts in the simulation exercise because of the pre-set limitation in the allowable numbers of building zones in ESP-r. It is shown in Fig. 6.9. Lobby, canteen, gym room and office were integrated into one building model, and the remaining zones were built in another one.

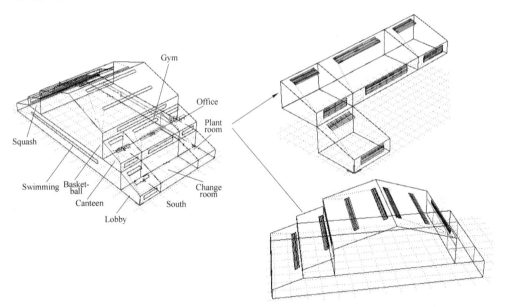

Fig. 6.9 Split of the sport center model into two parts

For the case study of Hong Kong, the TMY data was available in the weather database of the ESP-r platform. And for the year-sound simulation in the case of Beijing, the TMY weather data of Beijing was imported into the weather module of ESP-r. The inlet temperature of cold feed water was assumed the same as the ambient temperature in every time-step of the year-round building energy simulation.

The optical properties of the absorptive glazing were set the same as those covered in the validation exercise. For the IGU system with a sealed air layer, the glazing system was made composed of three layers, each given the equivalent optical properties of clear glazing, air layer and low-E glazing. The software Window developed at the Lawrence Berkeley National Laboratory was employed for the calculation of the optical properties of the glazing systems. The angle-dependent optical properties of the IGU system are given in Table 6.10.

Table 6.10 Variation of optical properties of IGU with incident angle

Incident angle	0	10	20	30	40	50	60	70	80
Tsol	0.306	0.308	0.304	0.297	0.289	0.273	0.239	0.173	0.08
Abs 1	0.343	0.346	0.352	0.356	0.355	0.355	0.361	0.357	0.282
Abs 2	0.030	0.030	0.030	0.031	0.031	0.032	0.031	0.027	0.020
Tvis	0.669	0.673	0.663	0.651	0.634	0.6	0.524	0.38	0.175
Rsol-f	0.321	0.316	0.314	0.316	0.325	0.34	0.37	0.443	0.619
Rsol-b	0.328	0.323	0.32	0.318	0.32	0.329	0.357	0.434	0.606

Note: 1. Tsol, Abs, Tvis and Rsol mean solar transmittance, solar absorptivity, visible light transmittance and solar reflectance respectively.

2. f and b refer to front and back sides respectively.

The direction of the transmitted solar radiation was required to be changed in the simulation run concerning liquid flow window system. The path of solar transmission was changed to make it reaching the inner glazing through the water layer. And thus, a part of the solar thermal energy was made absorbed by the flowing liquid stream. It was the same as in the validation exercise. Also, the thermodynamic properties, including solar absorption coefficient, density, thermal conductivity and specific heat capacity of liquid layer were imposed to replace the default properties of air. The convective heat transfer coefficients decided by the FORTRAN calculation at the glazing surfaces were also assigned because the heat transfer coefficients were determined by default in ESP-r by taking the fluid in cavity as air. And the average values of the heat transfer coefficients calculated with the self-developed computer program were used in these whole building energy simulation runs.

The designation of operational details of occupant, lighting and equipment as the three heat sources of the indoor environment was compulsory in ESP-r simulation. The occupant, lighting, equipment density as well as the required fresh air rates of all the function rooms are listed in Table 6.11. There were totally 2 badminton courts, and the maximum number of occupants for each of them was 4. Similarly, two playgrounds could be placed in the basketball game hall and the maximum number of players was 48 with four teams. Equipment loads were considered only in office. For canteen, heat gain of food was taken as 20W, half of it was sensible heat and the remaining part was latent heat.

Table 6.11 Occupants, lighting, equipment power densities and fresh air requirements of the function rooms at the sport center

Fuction room	Occupant density /$m^2 \cdot person^{-1}$	Lighting power density /$W \cdot m^{-2}$	Equipment power density /$W \cdot m^{-2}$	Minimum outdoor air /$L \cdot (s \cdot person)^{-1}$
Basketball hall	48	13	—	13
Badminton hall	8	17	—	13
Indoor swimming pool	3	15	—	13
Gym room	3	13	—	13
Canteen	1	17	—	10
Office	8	17	25	8
Main lobby	8	14	—	8

The total floor area for the seven function rooms and their corresponding heat release decided based on floor area are given in Table 6.12. For the heat source of occupant, it was determined based on the number of occupants and the release of sensible and latent heats for each of them. In office and lobby, the releases of sensible and latent heat were 75W and 60W per person respectively; the release of sensible and latent heat of people in canteen was determined based on low-grade activity similar to shop, and the corresponding values were 85W and 87W. However, for those buildings with intense activities like swimming pool, badminton and basketball game hall, the releases of sensible and latent heat were 152W and 255W per person. The values of sensible and latent heat release were the average values under the temperatures of 21℃ and 25℃ as the room temperature was controlled between 21℃ and 25℃.

Table 6.12 Requirements of power and fresh air rate of each function room at the sport center

Function room	Total floor area/m^2	Occupant /person	Lighting power /W	Equipment power/W	Minimum outdoor air/$L \cdot s^{-1}$
Basketball hall	900	48	11700	—	3900
Badminton hall	400	8	6800	—	104
Indoor swimming pool	1500	500	22500	—	6500
Gym room	200	67	2600	—	858
Canteen	100	100	1700	2000	1000
Office	100	13	1700	2500	96
Main lobby	100	13	1400	—	96

All function rooms of the sport center were not fully occupied throughout the day. The operation hours of the buildings were also different during weekday and weekend. The hourly varied operational schedules of occupants and lightings are given in Table 6.13 and Table 6.14, respectively. Operation schedule of buildings used for sports including gym room, badminton and basketball halls are the same, and only the operation details of the gym room are introduced here. For office, it was closed during Saturday afternoon and the whole Sunday. Lobby was opened everyday and with the same operation pattern throughout the seven days. The number of users of canteen was greater during weekend as compared to weekday. For the operation of artificial lighting, the lighting demand remains the same for all days in the cases of lobby and canteen. The closed office has no lighting demand and the lighting requirement of the sport rooms were the same in Saturday and Sunday.

Table 6.13 Occupancy schedules of function rooms

Time	Office		Main lobby	Gym room		Canteen		
	Weekday	Weekends	Everyday	Weekday	Weekends	Weekday	Saturday	Sunday
1	0	0	0	0	0	0	0	0
2	0	0	0	0	0	0	0	0
3	0	0	0	0	0	0	0	0
4	0	0	0	0	0	0	0	0
5	0	0	0	0	0	0	0	0
6	0	0	0	0	0	0	0	0
7	0.1	0.1	0.9	0.3	0.3	0.5	0.4	0.4
8	0.7	0.4	0.9	0.4	0.4	0.7	0.6	0.6
9	0.9	0.7	0.9	0.5	0.5	0.7	0.7	0.7
10	0.9	0.7	0.9	0.5	0.7	0.5	0.6	0.8
11	0.9	0.7	0.9	0.5	0.8	0.5	0.6	0.8
12	0.5	0.7	0.9	0.3	0.7	0.9	0.8	0.9
13	0.5	0.7	0.9	0.3	0.7	0.9	0.9	0.9
14	0.9	0	0.9	0.3	0.8	0.8	0.8	0.9
15	0.9	0	0.9	0.4	0.8	0.2	0.2	0.7
16	0.9	0	0.9	0.5	0.8	0.2	0.2	0.5
17	0.7	0	0.9	0.5	0.8	0.3	0.3	0.4
18	0.3	0	0.9	0.6	0.8	0.6	0.6	0.6
19	0.1	0	0.9	0.8	0.8	0.9	0.9	0.9
20	0.1	0	0.9	0.8	0.8	0.9	0.9	0.9
21	0.1	0	0.9	0.8	0.7	0.8	0.8	0.8
22	0.1	0	0.9	0.5	0.5	0.3	0.3	0.3
23	0	0	0	0	0	0.1	0.1	0.1
24	0	0	0	0	0	0	0	0

Table 6.14 Lighting schedules of the function rooms

Time	Office		Main lobby	Gym room		Canteen
	Weekday	Weekend	Everyday	Weekday	Weekend	Everyday
1	0.05	0.05	0.05	0.05	0.05	0.1
2	0.05	0.05	0.05	0.05	0.05	0.1
3	0.05	0.05	0.05	0.05	0.05	0.1
4	0.05	0.05	0.05	0.05	0.05	0.1
5	0.05	0.05	0.05	0.05	0.05	0.1
6	0.05	0.05	0.05	0.05	0.05	0.5
7	0.3	0.3	0.9	0.8	0.8	0.9
8	0.8	0.8	0.9	0.9	0.9	0.9
9	0.9	0.9	0.9	0.9	0.9	0.9
10	0.9	0.9	0.9	0.9	0.9	0.9
11	0.9	0.9	0.9	0.9	0.9	0.9
12	0.9	0.9	0.9	0.9	0.9	0.9
13	0.9	0.9	0.9	0.9	0.9	0.9
14	0.9	0.8	0.9	0.9	0.9	0.9
15	0.9	0.6	0.9	0.9	0.9	0.9
16	0.9	0.5	0.9	0.9	0.9	0.9
17	0.9	0.5	0.9	0.9	0.9	0.9
18	0.9	0.3	0.9	0.9	0.9	0.9
19	0.8	0.3	0.9	0.9	0.9	0.9
20	0.5	0.1	0.9	0.9	0.9	0.9
21	0.3	0.1	0.9	0.9	0.9	0.9
22	0.1	0.05	0.9	0.9	0.9	0.9
23	0.05	0.05	0.05	0.05	0.05	0.7
24	0.05	0.05	0.05	0.05	0.05	0.1

Some part of the ESP-r source codes were modified because of the replacement of the small window units with large window. The heat transfer mechanism between the glazing surface and the liquid layer of large window was the same as that of small window, but the area was multiplied. And the thermal extraction of the flowing liquid layer was thus multiplied with large window; it was the sum of thermal extraction of all the small window units. This was achieved by multiplying the mass flow rate calculated for a single piece of small window with the number of small window units in the large window. Thus, the liquid temperature at the outlet of the window cavity was maintained the same as that of small window because both heat gains and heat losses of the window system were multiplied with the number of windows. And the same heat exchanger used for a single piece of window was used to maintain the same system performance. The water heat gains of cold feed water in the heat exchanger were calculated by a control function and written into a readable file. The dimension of the heat exchanger and the cold feed water velocity were also defined in this control function.

6.3.3 Energy Saving Performance of Liquid Flow Window

The room heating and cooling loads were affected by the flowing liquid layer. Thermal energy was extracted by the liquid layer from both solar radiation and the two glazing panes. Room cooling loads of the sport center with Low-e glazing and liquid flow window in Hong Kong and Beijing are given in Table 6.15.

Table 6.15 Cooling loads of all the function rooms in Hong Kong and Beijing with different window systems (kW · h)

Function rooms	Hong Kong		Beijing	
	Low-e	Liquid glazing	Low-e	Liquid glazing
Indoor swimming pool	228046	228162	121314	118507
Lobby	13223	12981	6731	5915
Canteen	44073	38168	23111	18673
Gym room	51243	46957	28283	23518
Office	21909	20409	14094	11283
Basketball hall	144958	142383	75114	73016
Badminton hall	45379	46270	25221	25719
Sum	548831	535330	293868	276631

Cooling loads of Beijing were smaller than those of Hong Kong with large difference. And the adoption of liquid flow window was good for the cooling loads reduction except for the indoor swimming pool and badminton hall in Hong Kong. The use of liquid flow window in indoor swimming pool and badminton hall even resulted in higher cooling loads in Hong Kong, because these two buildings had north facing vertical windows and the absorbed solar energy during daytime was small. The liquid flow velocity was thus of small value and consequently the contribution of thermal extraction was not significant, the heat transfer between the window system and the indoor environment might even be enhanced when the flow was extremely slow. Thermal storage of the liquid flow window also resulted in the increase of cooling loads in the early morning.

For the main lobby, canteen, gym room and office under the similar condition, cooling loads could be reduced by 1.8%, 13.4%, 8.36% and 6.8% respectively in Hong Kong by using liquid flow window, and the corresponding reductions in Beijing were larger with differences of 12.11%, 19.2%, 16.85% and 19.94%. The larger difference in Beijing was caused by its stronger solar radiation and higher liquid flow velocity in the window cavity. Canteen benefited the most from the installation of liquid flow window in both Hong Kong and Beijing. Because the casual gain intensity of canteen was larger than the other rooms and thermal extraction of the liquid layer from the indoor environment also resulted in a decrease in room cooling loads. Large amount of thermal absorption was good for increase in liquid flow velocity and it would further enhance thermal extraction as a consequence. The reduction in cooling loads of office was larger than lobby and gym room in Beijing, because the cooling demands during working hours, when the solar radiation was strong, was the major source of the cooling loads for office use in Beijing.

However, the condition in Hong Kong was different with a larger reduction in gym room as compared to office, because the ambient temperature was high in Hong Kong even during the nighttime and cooling was required almost throughout the year, thus buildings with larger glazing area and cooling demands could benefit more from liquid flow window. The smaller decrease in the cooling loads of the main lobby as compared to the other buildings was caused by its small casual gain intensity, and consequently the small thermal extraction capability from the indoor environment. The impact of thermal storage became more important when the thermal extraction capability was small. The smaller decrease in the cooling loads of basketball game hall was caused by two reasons. Firstly, the small casual gain intensity; and secondly, the low thermal extraction capability of the liquid layer in the north facing windows.

The thermal extraction of the liquid layer resulted in the decrease in window surface temperature, and consequently an increase in the heating loads during the winter season. The heating loads in the subtropical city Hong Kong was small. But the weather condition of Beijing was different; it had large heating demands during the winter season and the nighttime of early spring and late autumn. Heating loads of the sport center in Hong Kong and Beijing with different window systems are listed in Table 6.16.

Heating was required in Jan., Feb. and Dec. for almost all the function rooms in Hong Kong except for office, which had almost no heating demand. The zero heating loads of office were caused by its short operation period and the large amount of internal heat sources from occupants, lighting and equipment. The increase in the heating loads of canteen was also evident because of the effective thermal extraction of the liquid layer. The heating loads of lobby were large because the thermal release from occupants and lighting was small and the room temperature was low. The increase in the heating loads

Table 6.16 Heating loads of all the function rooms in Hong Kong and Beijing with different window systems   (kW·h)

Function rooms	Hong Kong		Beijing	
	Low-e	Liquid glazing	Low-e	Liquid glazing
Indoor swimming pool	2803	3504	185291	203388
Lobby	130	160	7219	8289
Canteen	56	131	18473	22151
Gym room	0	2	11427	20623
Office	0	0	219	722
Basketball hall	3443	4417	129026	138186
Badminton hall	3	46	6078	6719
Sum	6435	8260	357733	400078

was 1825kW·h, and the reduction in the cooling loads was about 13502kW·h in Hong Kong.

However, the heating loads in Beijing were much larger and heating was required in most of the time from Oct. to Apr. (the next year), the exception of office building with smaller heating requirement was caused by its shorter air-conditioned period and the large internal heat source. The increase in the heating loads with liquid flow window was much larger as compared to Hong Kong because of its lower ambient temperature. Similar to the results in Hong Kong, the heating loads of the indoor swimming pool and the basketball hall were larger because of their larger space. The overall increase in the heating loads was even larger than the total reduction in the cooling loads with values of 42345kW·h and 17236kW·h.

Another contribution of the liquid flow window was the useful cold water heat gains except for the reduction in cooling loads. The total water heat gains of all the function rooms in Hong Kong and Beijing are given in Table 6.17. The water heat gains were partly from the absorption of solar thermal energy and partly from the thermal extraction of the liquid layer from the glazing panes. For buildings like indoor swimming pool, badminton hall and basketball game hall, half of whose windows were north facing; the liquid flow velocity was small, and thus the water heat gain was mainly affected by the solar condition. The stronger solar radiation in Beijing caused larger amount of water heat gains in these three function rooms in Beijing as compared to Hong Kong. But the water heat gains of canteen, gym house and office in Hong Kong were larger than those in Beijing. It was caused by the larger amount of solar thermal absorption and casual gain intensities as well as the higher ambient temperature in Hong Kong. The effect of liquid flow velocity was dominant under this circumstance and the liquid flow velocity in Hong Kong was larger with water as working fluid, thus its water heat gains were of larger amounts. However, the condition of lobby was different because its casual gain intensity was small and solar radiation was not as strong as that in Beijing.

Table 6.17 Water heat gains of liquid flow windows in
all the function rooms in Hong Kong and Beijing   (kW · h)

Function rooms	Water heat gains	
	Hong Kong	Beijing
Main lobby	3474	3505
Canteen	4709	4419
Gym room	9368	8471
Office	5563	4180
Indoor swimming pool	8326	11911
Basketball game hall	32179	42684
Badminton game hall	8857	12080
Sum	72476	87250

For south facing rooms including main lobby, canteen, gym room and office in Hong Kong, the water heat gains of office with unit glazing area was of the highest value. It was caused by the different control methods of the office; the room temperature was high during the period without air-conditioning operation and large amount of thermal energy was absorbed by the flowing liquid layer. However, Beijing could not benefit from this because the ambient temperature was much lower than that of Hong Kong, especially during the nighttime. Water heat gains of canteen were large in both Hong Kong and Beijing because of its high casual gain intensity and thermal extraction capability of the liquid layer as mentioned before. This same reason resulted in a larger water heat gain density of gym room as compared to lobby. The sum of water heat gains of all the function rooms were 72476kW · h and 87250kW · h respectively in Hong Kong and Beijing.

6.4 Payback Period Analysis

The total extra investment, energy input and carbon emission of windows installed in the sport center are summarized in Table 6.18.

Table 6.18 Investment, energy input and carbon emission for window systems in the sport center in Hong Kong and Beijing

City	Investment/CNY		Embodied energy/MW·h		Equivalent carbon emission	
	Stainless steel frame	Wood frame	Stainless steel frame	Wood frame	Stainless steel frame	Wood frame
Hong Kong	184000	83200	92800	69600	37200	16800
Beijing	454000	353200	176000	152800	44000	24000

Savings in electricity caused by the variation of air-conditioning load and the useful water heat gain in Hong Kong and Beijing were 75929kW·h and 82764kW·h respectively. It was calculated by taking COP of the air-conditioning system as 3.5 and 4.5 for cooling and heating, and thermal efficiency of electrical water heater as 0.99. The saving in the form of primary energy should be 227786kW·h and 248291kW·h respectively considering an electrical efficiency of 33%. And the corresponding saving of money were 75929 HKD and 64638 CNY by taking prices of electricity as 1 HKD/kW·h and 0.781 CNY/kW·h in Hong Kong and Beijing. The cost saving in Hong Kong was then 60743 CNY considering an exchange rate of 0.8 of HKD to CNY. Since the simulations were completed based on the weather condition of Hong Kong and Beijing, carbon emission rates of 0.7kg CO_2-eq/kW·h and 0.891kg CO_2-eq/kW·h for electricity generation were used. And the reduction in carbon emission should be 53150kg CO_2-eq and 73743kg CO_2-eq respectively.

The electricity saving could be reduced to 21242kW·h and 16893kW·h in Hong Kong and Beijing if heat pump water heater with COP of 4.2 was used for water heating, and the corresponding cost saving was decreased to 16994 CNY and 13193 CNY respectively. The corresponding saving in the form of primary energy were then 63727kW·h and 50678kW·h. And the reductions in carbon emission were 14869kg CO_2-eq and 15051kg CO_2-eq.

The calculated CPBT, EPBT and GPBT are given in Table 6.19 to Table 6.21. The payback period varied with frame materials and the water heating system selection. As expected, payback period was shorter with wooden frame as compared to that with stainless steel frame. The CPBT in Hong Kong was well below 3 years while electrical water heater was used, and it was less than 10.8 years even the heat pump water heating system with higher efficiency was considered. The cost payback time in Hong Kong was not more than 4.9 years with the wooden frame. But the payback period in Beijing was much longer because of the high cost of anti-freezing liquid. And it was as long as about 34.4 years when the heat pump water heater was considered. However, it was well below 7 years when the electrical water heating system was used. The payback time should be

Table 6.19 CPBT of liquid flow window with different frame materials and water heaters in Hong Kong and Beijing

Heating device	Frame material	Stainless steel frame		Wood frame	
	City	Hong Kong	Beijing	Hong Kong	Beijing
	Investment/CNY	184000	454000	83200	353200
Electrical heater	Cost saving/CNY	60743	64638	60743	64638
	CPBT/year	3.0	7.0	1.4	5.5
Heat pump	Cost saving/CNY	16994	13193	16994	13193
	CPBT/year	10.8	34.4	4.9	26.8

Table 6.20 EPBT of liquid flow window with different frame materials and water heaters in Hong Kong and Beijing

Heating device	Frame material	Stainless steel frame		Wood frame	
	City	Hong Kong	Beijing	Hong Kong	Beijing
	Embodied energy/kW·h	92800	176000	69600	152800
Electrical heater	Energy saving/kW·h	227786	248291	227786	248291
	EPBT/year	0.41	0.71	0.31	0.62
Heat pump	Energy saving/kW·h	63727	50678	63727	50678
	EPBT/year	1.46	3.47	1.09	3.02

Table 6.21 GPBT of liquid flow window with different frame materials and water heaters in Hong Kong and Beijing

Heating device	Frame material	Stainless steel frame		Wood frame	
	City	Hong Kong	Beijing	Hong Kong	Beijing
	Carbon emission/kg CO_2-eq	37200	44000	16800	24000
Electrical heater	Emission reduction/kg CO_2-eq	53150	73743	53150	73743
	GPBT/year	0.70	0.60	0.32	0.33
Heat pump	Emission reduction/kg CO_2-eq	14869	15051	14869	15051
	GPBT/year	2.50	2.92	1.13	1.59

shorter if the reduction in the installation of traditional water heating system caused by the use of liquid flow window was considered.

Energy and green-house gas payback times were less than 3.5 years even with stainless steel window frame and heat pump for water heating. And energy payback time could be less than 1 year by using electrical water heater. Similarly, GPBT was less than 1 year with electrical water heater, and the longer payback time in Hong Kong was caused by the higher carbon emission rate for electricity generation in Beijing. Then the contribution to carbon emission is much more significant.

In both Hong Kong and Beijing, the cost payback time was much longer than the energy and green house gas payback time, caused by the high cost of frame and pipe works. And the extraordinary long cost payback period in Beijing was related to the high price of anti-freezing liquid. For this reason, sensitivity analysis was conducted for the CPBT. In both Hong Kong and Beijing, a cost reduction of the accessory components by 50% was examined. Furthermore, the impacts of price and working mode of the anti-freezing liquid for the application in Beijing were studied through the following two cases:

(1) Case 1: Cost reduction of anti-freezing liquid by 50% and one replacement in the maintenance process.

(2) Case 2: Cost reduction of anti-freezing liquid by 50% and no replacement in the maintenance process.

With the reduction in the cost of accessory components, the CPBT in Hong Kong would be about 1.5 years and 0.6 years with steel and wooden frame structures when electrical water heating device was replaced. They were 5.4 years and 2.1 years when heat pump was taken as the traditional water heater. The payback periods in Beijing are given in Table 6.22. The CPBT was less than 3.6 years when the window system was in replace of electrical water heater, and it was less than 17.4 years even heat pump water heating device is considered. And the shorter payback period of Case 2 indicates that life time extension of working fluid is quite effective in shortening the cost payback time.

Table 6.22 CPBT of liquid flow window in Beijing with modified cases

Heating device	Frame material	Case 1	Case 2
Electrical heater	Steel frame	3.6	2.1
	Wooden frame	2.7	1.2
Heat pump	Steel frame	17.4	10.2
	Wooden frame	13.1	5.9

6.5 Summary

A comprehensive life cycle impact assessment from the economic, energy and environmental aspects was applied to a novel liquid flow window. It was completed taking the weather conditions of Hong Kong and Beijing. And the building energy simulation was carried out with ESP-r platform. The major findings are given below:

(1) Energy input and carbon emission for construction of liquid flow window are re-

sulted mainly from glazing and frame materials. And the contribution of frame structure and pipe work is dominant in the investment. Another significant cause of high cost in Beijing application is the use of anti-freezing liquid.

(2) South facing buildings and those buildings with larger casual gain intensities benefit more from liquid flow window because larger amount of thermal energy can be extracted by the liquid layer. Thus users can benefit from energy saving in both the reduction in the cooling loads and the increase in the useful water heat gains. However, the thermal extraction of the liquid layer also results in large increase in the room heating loads; especially in Beijing.

(3) South facing buildings including canteen, gym and office have larger useful water heat gains in Hong Kong as compared to Beijing because of the large liquid flow velocity caused by strong solar radiation, large casual gain intensity and small water viscosity. But for north facing buildings, and building with quite small casual gain intensity like lobby, the water heat gains in Beijing is larger as compared to Hong Kong since the main source of the water heat gain is solar thermal energy absorption with a small liquid flow velocity.

(4) The payback time is affected by the material selection and the thermal efficiency of the water heating systems. Shorter payback time could be obtained with wooden frame. And the payback period is short with CPBT in Hong Kong less than 3 years; EPBT and GPBT within 1 year while electrical heater was used for water heating. However, the payback periods are much longer if heat pump water heater with a larger thermal efficiency is used for water heating. The CPBT could be as long as 10.8 and 34.4 years respectively in Hong Kong and Beijing, while the EPBT and GPBT were below 3.5 and 3 years.

(5) Cost payback time can be further reduced by adopting compact and better design of frame structure and pipe works as well as using anti-freezing liquid with lower price and longer life time. The longest cost payback time in Hong Kong can be reduced from 10.8 to 5.4 years by reducing the cost of frame structures by half. And the payback time in Beijing can be reduced to the range from 1.2 years to 10.2 years by reducing the price of frame structure and anti-freezing liquid by half and increasing the lifetime of anti-freezing liquid to the same lifetime of the window system.

7 Summary and Major Conclusions

7.1 Research Work Completed

The main focuses of this study are thermal performance evaluations of the liquid flow window system to optimize its schematic design and to improve the energy saving performance under various application conditions, ie. different climate conditions and different building types. The economic, energy and environment impacts assessment is also included. To achieve the above, it was completed by employing combined experimental and numerical approaches.

A literature review was firstly carried out regarding the development of advanced glazing technologies and their applications in buildings. The glazing techniques have gone through the development from single to multi-layer systems with the improved thermal performance. And liquid flow glazing technology introduced in this book is one of the novel multi glazing systems.

Secondly, experimental evaluation of thermal performance of the liquid flow window with working fluids of water and anti-freezing liquid, which was aqueous solution of propylene glycol at 40% concentration, was introduced. The glazing material used was absorptive glazing with dimensions of 1.2m(H)×0.8m(W) and thickness of 12mm. It was thick enough to resist the possible deformation or damage caused by large liquid pressure. Window was installed on the south facing wall of an outdoor test chamber with equal length, width and height of 3m. In both experiments, solar radiation level was measured with pyranometer; glazing surface temperatures, fluid temperatures in the heat exchanger and environment temperatures were detected with T type thermocouple. The measured data was used for thermal performance analysis and numerical approach validation. Thermal performance analysis was completed by comparing thermal transmission through the window and thermal collection capacity. Cold water heat gain was addressed based on its temperature increase and the measured supply velocity. And thermal efficiency was determined as a ratio of water heat gain to solar radiation incident on the window surface.

Model validation tests have been completed successfully on both the modified FOR-

TRAN program and the ESP-r software for both water flow and anti-freezing liquid flow experiments. Measured solar radiation on the horizontal surface, room and ambient temperatures, as well as temperature and velocity of cold feed water were used as inputs in the simulation study. And the predicted glazing surface temperatures, hot and cold streams temperatures in the heat exchanger were compared with the experimental results for the purpose of modelling quality illustration. Uncertainty analysis was completed to find the potential deviation between simulation and experiment results caused by measurement errors and uncertainties of inputs. Error bands were determined and the validation study was finally taken as successful when the deviations between simulation and experimental results located well within the error bands. Successful validation with the modified computer program was to provide basis for the year-round performance prediction in schematic and application optimization. With the successful validation in ESP-r, whole building energy simulation could be conducted on this innovative technology.

Performance evaluation regarding schematic optimization was introduced considering the variation in liquid layer thickness, pipework design and glazing hight to width ratio. The thickness of the liquid layer varied from 30mm to 20mm, 15mm and 10mm. The proper setting of pipe work was investigated by comparing system thermal performance with and without distribution headers. Covariation of cavity space and connection pipe size was considered with pipe diameter in the range from 10mm to 30mm. Height to width ratio of the glazing varied in the range from 3.2 to 0.24 with different glazing areas of $1.6m^2$, $0.96m^2$ and $0.8m^2$. The above were conducted via numerical approach and so mathematical models were developed for modelling the solar distribution on the tilted surface, heat transfer in the glazing panes and the liquid layer, liquid flow in the window cavity and heat transfer in the heat exchanger. Performance of the system was compared from the aspects of water heat gain, heat flow through the window, and energy saving impact considering the balance between the saving in water heater and the consumption of air-conditioning system.

To further improve the system performance, PCM was considered to be added to the double pipe heat exchanger and to make it into a triple-pipe heat exchanger. Two designs were proposed by placing the PCM layer to different positions. In the first case, it was placed in the annular space between hot and cold streams; in the second design, it was added to the outermost layer of the original double-pipe heat exchanger. In this study, mathematical model for heat transfer in the PCM layer was also developed and the model was validated by comparing with published experimental data. It was expected to achieve the purposes of load shifting and heat transfer enhancement by using PCM in the heat exchanger. Numerical evaluation of system performance was conducted during typical summer and winter weeks as extracted from the TMY weather data set of Hong Kong. The system performance was evaluated from several aspects, which include the variation in useful cold water heat gains and room heat exchanges. Results from double-pipe heat exchanger were compared to those from the triple-pipe heat exchanger.

To expand its application in cold climates, passive freezing prevention approach was adopted. And the energy saving performance of liquid flow window was evaluated in seven different climate zones of China numerically with the validated FORTRAN program. For those areas with cold winter, anti-freezing liquid was used, and its concentration was determined by the local climate condition. The impact of anti-freezing liquid concentration on system thermal performance was firstly investigated based on the weather condition of Guangzhou, followed by the year-round simulations in the different climate zones. Besides, the effectiveness of water drainage in winter season was also investigated.

Furthermore, LCA was conducted from the aspects of economic, energy and environmental impact, CPBT, EPBT and GPBT were determined. Investment, embodied energy and carbon emission for system construction, transportation, maintenance and disposal were brought into consideration, and also energy saving came from the useful water heat gain of cold feed water and the reduction in air-conditioning load. For window system construction, different frame materials including stainless steel and wood were considered. This study was completed with findings on the application of liquid flow window in a sport center. Year-round simulations were conducted in ESP-r with two glazing systems: the liquid flow window and the IGU(6mm clear +12mm air + 6mm low-e) system, under the weather conditions of Hong Kong and Beijing. Payback periods regarding economic, energy and environmental impact were compared, and the differences in payback periods with variation of frame materials and water heating systems were also evaluated.

7.2 Findings and Conclusions

Based on the above mentioned research works, some of the findings are concluded here:

(1) With water and anti-freezing liquid as working fluids, average thermal efficiencies during the testing days were 8.03% and 7.2% respectively. It was affected by both weather parameters and fluid properties. Expose of cold water feed pipe to direct solar

radiation would lead to an increase in feed water inlet temperature and consequently a decrease in useful water heat gain and system thermal efficiency. However, from the points of energy harvesting and utilization, the water stream in an uninsulated pipe could be heated by the solar radiation directly before delivering to the hot water system. The contribution of temperature increase to electricity saving of hot water system could also be promising in the summer season.

(2) Variation in the thickness of the liquid layer in the window cavity affected the absorption of solar thermal energy and the thermal resistance along the flow path. With a thinner liquid layer, liquid temperature increased faster and the corresponding useful water heat gain was slightly larger. However, the reduction in liquid layer thickness would also result in an increase in thermal transmission via the window system because of the decreased thermal resistance. And the impacts on heat transmission via the window system were found to be larger as compared to that on the useful water heat gain. In the present case study with glazing dimensions of 1.2m(H)×0.8m(W)×0.12m(D), the reduction of liquid layer from 30mm to 10mm was good for system thermal efficiency improvement. However, with a liquid layer of 15mm or 10mm, the improvement in thermal efficiency was fairly small while the increases in the room heat gain and heat loss were large. And a 20mm water layer was a better choice in the current study.

The system thermal performance was found also affected by the setting of distribution headers and the flowing pipe size. Water circulation could be improved by eliminating the distribution headers and the temperature distribution at the inner glazing surface was more uniform. It contributed to the increase of yearly water heat gain by about 5%, and a minor drop in room heat gain. Besides, the connecting-pipe size showed significant influence on its thermal performance. The increase in pipe diameter resulted in the reduction of friction loss, thus water circulation and heat transfer coefficient could be improved. The recommended pipe diameter is 20~25mm, with comparative high thermal efficiency of 13.58% to 13.93%.

With a constant glazing area, reducing of height to width ratio contributed to performance improvement because of the reduced thermal resistance along the flow path and the increased heat transfer area in the heat exchanger. However, the space occupied by the heat exchanger could be larger with a longer heat exchanger sealed in the window frame. And thus the thermal effect caused by unit window area should be different from that of unit glazing area. The reduction of height to width ratio to about 0.4 was good for system performance improvement concluded from the case studies with glazing areas of 1.6m^2, 0.96m^2 and 0.8m^2.

(3) Both validation tests with the modified FORTRAN program and ESP-r were successful. Year-round simulations carried out with the validated FORTRAN program in different climate conditions indicated that promising system thermal performance could be obtained under strong solar radiation; even the ambient temperature was not high enough. Electricity consumption of the water heating system could be reduced greatly because of the useful water heat gain with the liquid flow window, and this was especially true in Region V and Region VI as concluded from the results of Kunming and Lhasa. In Shenyang located in Region I, the overall contribution of liquid flow window was not as advantageous as in the other cities because of its low ambient temperature.

The system performance was highly affected by the anti-freezing liquid concentration. System thermal efficiency was lower with anti-freezing liquid as comparing to that with water under the same working condition. However, this difference could be minimized under very strong solar radiation. Replacement of anti-freezing liquid with water from Mar. to Oct. in Region I, II, VI and VII, which have freezing protection demand, contributed to system performance improvement. The improvements were especially high in Region I and Region VII with very cold winter. This preference was obtained without considering the extra requirement on operation and maintenance for liquid replacement.

(4) By applying PCM to the outermost layer of the original double pipe heat exchanger, the peak water heat gain could be shifted to the off-work hours for residential application. And for office use, which had no warm water demand during off-work hours, the large amount of stored thermal energy would be released to the cold feed water in the early morning. Another benefit of this design was the enhanced heat transfer and larger amount of useful water heat gain as compared to the original double pipe heat exchanger. The increase in water heat gain was found larger during typical summer week than in typical winter week because of the larger solar intensity and the higher ambient temperature during summer season. This was the same for both office and residential buildings.

However, the system thermal performance was deteriorated when the PCM layer was located in the annular space between the hot and cold fluid streams. Thermal energy released to the cold water reduced greatly because of the larger thermal resistance of the PCM layer. And this would result in an increase in room heat gains because the return water was of higher temperature. The reduction in water heat gains and the increase in air-conditioning loads were observed for both office and residential uses under such condition.

(5) In the life cycle assessment, payback periods regarding cost, energy and carbon

for a liquid flow window with wooden frame was shorter than that with stainless steel frame because wood was much lighter and carbon emission for wood production was smaller. Frame structures and pipe works contributed to the major part of investment. And for the application in Beijing, another significant influential factor on the cost was the use of anti-freezing liquid. But glazing was the major cause of energy consumption and carbon emission. With electrical water heaters in the sport center, CPBT was less than 3 years in Hong Kong and EPBT and GPBT were less than 1 year in both Hong Kong and Beijing. The longer cost payback time in Beijing was caused by the high cost of anti-freezing liquid and its replacement during maintenance. The payback periods could be longer if heat pump water heater was adopted, especially for the cost payback time in Beijing, which could be as long as 34.4 years, but the short time of about 3.5 and 2 years were achieved for EPBT and GPBT.

The cost payback time was further reduced by decreasing the cost of accessory components and anti-freezing liquid and prolonging the life time of the anti-freezing liquid. By reducing the cost of accessory components by 50%, the maximum payback time in Hong Kong was reduced to 5.4 years. And the maximum payback period in Beijing could be around 10 years by cuting the price of anti-freezing liquid by half and prolonging its lifetime to the same as window system.

References

[1] J. Houghton. *Global Warming: The Complete Briefing*, 4th Edition [M]. Cambridge University Press, Cambridge, 2009.

[2] J. Hansen, R. Ruedy, M. Sato, et al. Global surface temperature change [J]. Reviews of geophysics, 2010, 48: 1-28.

[3] C. L. Quéré, R. J. Andres, T. Boden. The global carbon budget 1959-2011 [J]. Earth System Science Data, 2013, 5: 165-185.

[4] The World Wind Energy Association. 2014 Half-year Report [Z], 1-8, 2014.

[5] L. Pérez-Lombard, J. Ortiz, C. Pout. A review on buildings energy consumption information [J]. Energy and Buildings, 2008, 40: 394-398.

[6] X. G. Casals. Analysis of building energy regulation and certification in Europe: Their role, limitations and differences [J]. Energy and Buildings, 2006, 38: 381-392.

[7] Z. J. Ma, S. W. Wang. Building energy research in Hong Kong [J]. Renewable and Sustainable Energy Reviews, 2009, 13: 1870-1883.

[8] Y. Jiang, X. Yang. China building energy consumption situation and the problems exist in the energy conservation works [J]. China construction, 2006, 2: 12-17.

[9] M. F. Tang. Solar control for buildings [J]. Building and Environment, 2002, 37: 659-664.

[10] S. B. Sadineni, S. Madala, R. F. Boehm. Passive building energy savings: A review of building envelope [J]. Renewable and Sustainable Energy Review, 2011, 15: 3617-3631.

[11] T. T. Chow, C. Y. Li, Z. Lin. Innovative solar windows for cooling-demand climate [J]. Solar Energy Materials and Solar Cells, 2010, 94: 212-220.

[12] J. Mohelnikova. Materials for reflective coatings of window glass applications [J]. Construction and Building Materials, 2009, 23: 1993-1998.

[13] H. Arsenault, M. Hébert, M. C. Dubois. Effects of glazing color type on perception of daylight quality, arousal, and switch-on patterns of electric light in office rooms [J]. Building and Environment, 2012, 56: 223-231.

[14] A. Jonsson, A. Roos. Visual and energy performance of switchable windows with antireflection coatings [J]. Solar Energy, 2010, 84: 1370-1375.

[15] T. Rosencrantz, H. Bülow-Hübea, B. Karlssona, A. Roos. Increased solar energy and daylight utilization using anti-reflective coatings in energy-efficient windows [J]. Solar Energy Materials and Solar Cells, 2005, 89: 249-260.

[16] B. P. Jelle, A. Hynd, A. Gustavsen, et al. Fenestration of today and tomorrow: A state-of-the-art review and future research opportunities [J]. Solar Energy Materials and Solar Cells, 2012, 96: 1-28.

[17] C. G. Granqvist. Electrochromism and smart window design [J]. Solid State Ionlcs, 1992, 53 (56): 479-489.

[18] A. M. Nilsson, A. Roos. Evaluation of optical and thermal properties of coatings for energy efficient windows [J]. Thin Solid Films, 2009, 517: 3173-3177.

[19] C. M. Lampert. Smart switchable glazing for solar energy and daylight control [J]. Solar Energy Materials and Solar Cells,1998,52:207-221.

[20] A. Piccolo, F. Simone. Effect of switchable glazing on discomfort glare from windows [J]. Building and Environment,2009,44:1171-1180.

[21] F. Gugliermetti, F. Bisegna. Visual and energy management of electrochromic windows in Mediterranean climate [J]. Building and Environment,2003,38:479 - 492.

[22] S. N. Alamri. The temperature behavior of smart windows under direct solar radiation [J]. Solar Energy Materials and Solar Cells,2009,93:1657-1662.

[23] A. Jonsson, A. Roos. Evaluation of control strategies for different smart window combinations using computer simulations [J]. Solar Energy,2010,84:1-9.

[24] K. Midtdal, B. P. Jelle. Self-cleaning glazing products: A state-of-the-art review and future research pathways [J]. Solar Energy Materials and Solar Cells,2013,109:126-141.

[25] E. Hammarberg, A. Roos. Antireflection treatment of low-emitting glazings for energy efficient windows with high visible transmittance [J]. Thin Solid Films,2003,442:222-226.

[26] A. Kaklauskas, E. K. Zavadskas, S. Raslanas, et al. Selection of low-e windows in retrofit of public buildings by applying multiple criteria method COPRAS: A Lithuanian case [J]. Energy and Buildings,2006,38:454-462.

[27] G. Kirankumar, S. Saboor, S. S. Vali, et al. Thermal and cost analysis of various air filled double glazed reflective windows for energy efficient buildings [J]. Journal of building engineering, 2020,28:1-14.

[28] K. A. R. Ismail, C. T. Salinas, J. R. Henriquez. Comparison between PCM filled glass windows and absorbing gas filled windows [J]. Energy and Buildings,2008,40:710-719.

[29] E. Cuce. Accurate and reliable U -value assessment of argon-filled double glazed windows: A numerical and experimental investigation [J]. Energy and Buildings,2018,171:100-106.

[30] H. J. Gläser, S. Ulrich. Condensation on the outdoor surface of window glazing——Calculation methods, key parameters and prevention with low-emissivity coatings [J]. Thin Solid Films, 2013,532:127-131.

[31] S. Y. Song, J. H. Job, M. S. Yeo, et al. Evaluation of inside surface condensation in double glazing window system with insulation spacer: A case study of residential complex [J]. Building and Environment,2007,42:940-950.

[32] E. Cuce, P. M. Cuce. Vacuum glazing for highly insulating windows: Recent developments and future prospects [J]. Renewable and sustainable energy reviews,2016,54:1345-1357.

[33] Y. P. Fang, T. J. Hyde, F. Arya, et al. Indium alloy- sealed vacuum glazing development and context [J]. Renewable and sustainable energy reviews,2014,37:480-501.

[34] A. Ghosh, B. Norton, A. Duffy. Measured thermal & daylight performance of an evacuated glazing using an outdoor test cell [J]. Applied Energy,2016,177:196-203.

[35] R. E. Collins, T. M. Simko. Current status of the science and technology of vacuum glazing [J]. Solar Energy,1998,62:189-213.

[36] Y. P. Fang, T. J. Hyde, F. Arya, et al. A novel building component hybrid vacuum glazing: a mod-

elling and experimental validation [J]. ASHRAE Trans, 2013, 119:430-441.

[37] S. Memon. Experimental measurement of hermetic edge seal's thermal conductivity for the thermal transmittance prediction of triple vacuum glazing [J]. Case studies in thermal engineering, 2017, 10:169-178.

[38] S. Memon, Y. P. Fang, P. C. Eames. The influence of low-temperature surface induction on evacuation, pump-out hole sealing and thermal performance of composite edge-sealed vacuum insulated glazing [J]. Renewable energy, 2019, 135:450-464.

[39] K. A. R Ismail, J. R. Henriquez. Parametric study on composite and PCM glass systems [J]. Energy Conversion and Management, 2002, 43:973-993.

[40] D. Li, Y. Y. Wu, B. C. Wang, et al. Optical and thermal performance of glazing units containing PCM in buildings: A review [J]. Construction and building materials, 2020, 233:117-327.

[41] F. Goia, M. Perino, V. Serra. Improving thermal comfort conditions by means of PCM glazing systems [J]. Energy and Buildings, 2013, 60:442-452.

[42] B. L. Gowreesunker, S. B. Stankovic, S. A. Tassou, et al. Experimental and numerical investigations of the optical and thermal aspects of a PCM-glazed unit [J]. Energy and Buildings, 2013, 61: 239-249.

[43] D. Li, Y. M. Zheng, Z. W. Li, et al. Optical properties of a liquid paraffin-filled double glazing unit [J]. Energy and Buildings, 2015, 108:381-386.

[44] Etzion Y, Erell E. Controlling the transmission of radiant energy through windows: A novel ventilated reversible glazing system [J]. Building and environment, 2000, 35:433-444.

[45] J. Zhou, Y. M. Chen. A review on applying ventilated double-skin facade to buildings in hot-summer and cold-winter zone in China [J]. Renewable and Sustainable Energy Reviews, 2010, 14: 1321-1328.

[46] S. Z. Movassag, K. Zamzamian. Numerical investigation on the thermal performance of double glazing air flow window with integrated blinds [J]. Renewable energy, 2020, 148:852-863.

[47] J. S. Carlos, H. Corvacho, P. D. Silva, et al. Heat recovery versus solar collection in a ventilated double window [J]. Applied Thermal Engineering, 2012, 37, 258-266.

[48] T. T. Chow, Z. Lin, W. He, et al. Use of ventilated solar screen window in warm climate [J]. Applied Thermal Engineering, 2006, 26:1910-1918.

[49] F. Gugliermetti, F. Bisegna. Saving energy in residential buildings: the use of fully reversible windows [J]. Energy, 2007, 32:1235-1247.

[50] T. T. Chow, Z. Z. Qiu, C. Y. Li. Potential application of 'see-through' solar cells in ventilated glazing in Hong Kong [J]. Solar Energy Materials and Solar Cells, 2009, 93:230-238.

[51] T. T. Chow, Z. Z. Qiu, C. Y. Li. Performance evaluation of PV ventilated glazing [R]. Eleventh International IBPSA Conference: Glasgow, Scotland. July 27-30, 2009.

[52] Y. P. Fang, F. Arya. Evacuated glazing with tempered glass [J]. Solar Energy, 2019, 183: 240-247.

[53] Y. P. Fang, T. J. Hyde, N. Hewitt. Predicted thermal performance of triple vacuum glazing [J]. Solar Energy, 2010, 84:2132-2139.

[54] Y. P. Fang, S. Memon, J. Q. Peng, et al. Ming. Solar thermal performance of two innova-tive configurations of airvacuum layered triple glazed windows [J]. Renewable Energy, 2020, 150: 167-175.

[55] M. X. Bao, X. G. Liu, J. Yang, et al. Novel hybrid vacuum/triple glazing units with pressure equalisation design [J]. Construction and Building Materials, 2014, 73: 645-651.

[56] S. H. Li, G. F. Sun, K. K Zou, et al. Experimental research on the dynamic thermal performance of anovel triple-pane building window filled with PCM [J]. Sustainable Cities and Society, 2016, 27: 15-22.

[57] S. H. Li, G. F. Sun, K. K. Zou, et al. Simulation research on the dynamic thermal performance of a novel tripleglazed window filled with PCM [J]. Sustainable Cities and Society, 2018, 40: 266-273.

[58] G. Michaux, R. Greffet, P. Salagnac, et al. Modelling of an airflow window and numerical investigation of its thermal performances by comparison to conventional double and triple-glazed windows [J]. Applied Energy, 2019, 242: 27-45.

[59] C. Zhang, J. B. Wang, X. H. Xu, et al. Modeling and thermal performance evaluation of a switchable triple glazing exhaust air window [J]. Applied Thermal Engineering, 2016, 92: 8-17.

[60] P. Sadooghi, N. P. Kherani. Influence of slat angle and low-emissive partitioning radiant energy veils on the thermal performance of multilayered windows for dynamic facades [J]. Renewable Energy, 2019, 143: 142-148.

[61] Y. P. Fang, T. J. Hyde, F. Arya, et al. Enhancing the thermal performance of triple vacuum glazing with low-emittance coatings [J]. Energy and Buildings, 2015, 97: 186-195.

[62] F. Goia, L. Bianco, Y. Cascone, et al. Experimental analysis of an advanced dyn-amic glazing prototype integrating PCM and thermotropiclayers [J]. Energy Procedia, 2014, 48: 1272-1281.

[63] F. Goia. Thermo-physical behaviour and energy performance assessment of PCM glazing system configurations: A numerical analysis [J]. Frontiers of Architectural Research, 2012, 1: 341-347.

[64] C. Y. Liu, G. J. Zhang, L. Dong, et al. Thermal performance of non-ventilated multilayer glazing facades filled with phase change material [J]. Solar Energy, 2019, 177: 464-470.

[65] C. Zhang, J. B. Wang, X. H. Xu. Analysis of the thermal performance of a switchable airflow window [R]. The 6th International Conference on Sustainable Development in Building and Environment, Chongqing, 25-27, October, 2013.

[66] C. Zhang, J. B. Wang, X. H. Xu, et al. Development of a simplified model of the switchable exhaust air insulation window [J]. Proceedings of the 2014 ASHRAE/IBPSA-USA Building Simulation Conference, 2014, 316-322.

[67] J. Wei, J. Zhao, Q. Chen. Energy performance of a dual airflow window under different climates [J]. Energy and buildings, 2010, 42(1): 111-122.

[68] J. Wei, J. Zhao, Q. Chen. Optimal design for a dual-airflow window for different climate regions in China [J]. Energy and buildings, 2010, 42(11): 2200-2205.

[69] F. A. Gonzalo, J. A. H. Ramos. Testing of water flow glazing in shallow geothermal systems [J]. Jocedia engineering, 2016, 161: 887-891.

[70] C. Y. Li, T. T. Chow. Water-filled double reflective window and its year-round performance [J]. Procedia Environmental Sciences, 2011, 11: 1039-1047.

[71] T. T. Chow, C. Y. Li, Z. Lin. Thermal characteristics of water-flow double-pane window [J]. International Journal of Thermal Sciences, 2011, 50: 140-148.

[72] T. T. Chow, C. Y. Li, Z. Lin. The function of solar absorbing window as water-heating device [J]. Building and Environment, 2011, 46: 955-960.

[73] T. T. Chow, C. Y. Li, J. A. Clarke. Numerical prediction of water-flow glazing performance with reflective coating [R]. 12th Conference of International Building Performances Simulation Association, Sydney, 14-16, November, 2012.

[74] T. T. Chow, Y. L. Lyu. Effect of design configurations on water flow window performance [J]. Solar energy, 2017, 155: 354-362.

[75] P. Sierra, J. A. Hernández. Solar heat gain coefficient of water flow glazings [J]. Energy and Buildings, 2017, 139: 133-145.

[76] Y. L. Lyu, T. T. Chow. Evaluation of influence of header design on flow characteristics in window cavity with CFD [J]. Energy Procedia, 2015, 78: 97-102.

[77] Y. L. Lyu, W. J. Liu, T. T. Chow, et al. Pipe-work optimization of water flow window [J]. Renewable energy, 2019, 139: 136-146.

[78] T. T. Chow, Y. L. Lyu. Numerical analysis on the advantage of using PCM heat exchanger in liquid-flow window [J]. Applied thermal engineering, 2017, 125: 1218-1227.

[79] Y. L. Lyu, T. T. Chow, J. L. Wang. Numerical prediction of thermal performance of liquid-flow window in different climates with anti-freeze [J]. Energy, 2018, 157: 412-423.

[80] T. Gil-Lopez, C. Gimenez-Molina. Environmental, economic and energy analysis of double glazing with a circulating water chamber in residential buildings [J]. Applied Energy, 2013, 101: 572-581.

[81] G. L. Tomas, G. M. Carmen. Influence of double glazing with a circulating water chamber on the thermal energy savings in buildings [J]. Energy and Building, 2013, 56: 56-65.

[82] G. L. Tomas, G. M. Carmen. Environmental, economic and energy analysis of double glazing with a circulating chamber in residential buildings [J]. Applied energy, 2013, 101: 572-581.

[83] C. Shen, X. T. Li. Solar heat gain reduction of double glazing window with cooling pipes embedded in venetian blinds by utilizing natural cooling [J]. Energy and buildings, 2016, 112: 173-183.

[84] C. Shen, X. T. Li. Thermal performance of double skin façade with built-in pipes utilizing evaporative cooling water in cooling season [J]. Solar energy, 2016, 137: 55-65.

[85] C. Shen, X. T. Li, S. Yan. Numerical study on energy efficiency and economy of a pipe-embedded glass envelope directly utilizing ground-source water for heating in diverse climates [J]. Energy conversion and management, 2017, 150: 878-889.

[86] C. Shen, X. T. Li. Energy saving potential of pipe-embedded building envelope utilizing low-temperature hot water in the heating season [J]. Energy and buildings, 2017, 138: 318-331.

[87] Y. L. Lyu, X. Wu, C. Y. Li, et al. Numerical analysis on the effectiveness of warm water supply in water flow window for room heating [J]. Solar energy, 2019, 177: 347-354.

[88] M. Ibrahim, E. Wurtz, J. Anger, et al. Experimental and numerical study on a novel temperature façade solar thermal collector to decrease the heating demands: A south-north pipe-embedded closed-water-loop system [J]. Solar energy, 2017, 147: 22-36.

[89] C. Y. Li, H. D. Tang. Evaluation on year-round performance of double-circulation waterflow window [J]. Renewable Energy, 2020, 150: 176-190.

[90] D. Gstoehl, J. Stopper, S. Bertsch, D. Schwarz. Fluidised glass façade elements for an active energy transmission control [R]. World engineers convention, Geneva, September 4-9, 2011.

[91] J. Stopper, F. Boeing, D. Gstoehl. Fluid glass façade elements: energy balance of an office space with a fluid glass façade [R]. Munich economic forum, Germany, 2013.

[92] A. Liebold, D. Gstoehl, D. Oppliger, S. Bertsch. Fluid glass-façade element for active solar control for high-rise buildings [R], 3th international high performance buildings conference at purdue, Purdue, America, July 14-17, 2014.

[93] C. Villasante, I. Hoyo, I. Pagola, et al. Solar active envelope module with an adjustable transmittance/absorptance [J]. Journal of Facade Design and Engineering, 2015, 3: 49-57.

[94] S-C. Kim, J-H. Yoon, H-M. Lee. Comparative experimental study on heating and cooling energy performance of spectrally selective glazing [J]. Solar Energy, 2017, 145: 78-89.

[95] M. G. Gomes, A. J. Santos, A. M. Rodrigues. Solar and visible optical properties of glazing systems with venetian blinds: Numerical, experimental and blind control study [J]. Building and Environment, 2014, 71: 47-59.

[96] R. Liang, M. Kent, R. Wilson, Y. Wu. Development of experimental methods for quantifying the human response to chromatic glazing [J]. Building and Environment, 2019, 147: 199-210.

[97] Y. Ajaji, P. André. Thermal comfort and visual comfort in an office building equipped with smart electrochromic glazing: an experimental study [J]. Energy Procedia, 2015, 78: 2464-2469.

[98] G. K. Kumar, S. Saboor, T. P. A. Babu. Experimental and Theoretical Studies of Window Glazing Materials of Green Energy Building in Indian Climatic Zones [J]. Energy Procedia, 2017, 109: 306-313.

[99] M. Chen, W. Zhang, L. Z. Xie, et al. Experimental and numerical evaluation of the crystalline silicon PV window under the climatic conditions in southwest China [J]. Energy, 2019, 183: 584-598.

[100] L. Giovannini, F. Favoino, A. Pellegrino, et al. Thermochromic glazing performance: From component experimental characterisation to whole building performance evaluation [J]. Applied Energy, 2019, 251: 113-135.

[101] A. Piccolo, C. Marino, A. Nucara, et al. Energy performance of an electrochromic switchable glazing: Experimental and computational assessments [J]. Energy and Buildings, 2018, 165: 390-398.

[102] C. Zhang, W. J. Gang, J. B. Wang, et al. Numerical and experimental study on the thermal performance improvement of a triple glazed window by utilizing low-grade exhaust air [J]. Energy, 2019, 167: 1132-1143.

[103] E. Cuce, P. M. Cuce. Optimised performance of a thermally resistive PV glazing technology: An

experimental validation [J]. Energy Reports,2019,5:1185-1195.

[104] R. Hart. Numerical and experimental validation for the thermal transmittance of windows with cellular shades [J]. Energy and Buildings,2018,166:358-371.

[105] P. Ma,L. S. Wang,N. Guo. Maximum window-to-wall ratio of a thermally automous building as a function of envelope U-value and ambient temperature amplitude [J]. Applied Energy,2015, 146:84-91.

[106] F. Goia. Search for the optimal window-to-wall ratio in office buildings in different European climates and implications on total energy saving potential [J]. Solar Energy,2016,132:467-492.

[107] I. Pérez-Grande, J. Meseguer, G. Alonso. Influence of glass properties on the performance of double-glazed facades [J]. Applied Thermal Engineering,2005,25:3163-3175.

[108] M. L. Persson,A. Roos,M. Wall. Influence of window size on the energy balance of low energy houses [J]. Energy and Buildings,2006,38:181-188.

[109] T. T. Chow, C. Y. Li. Liquid-filled solar glazing design for buoyant water-flow [J]. Building and Environment,2013,60:45-55.

[110] M. Arici,M. Kan. An investigation of flow and conjugate heat transfer in multiple pane windows with respect to gap width,emissivity and gas filling [J]. Renewable energy,2015,75:249-256.

[111] L. J. Claros-Marfil,J. F. Padial,B. Lauret. A new and inexpensive open source data acquisition and controller for solar research: Application to a water-flow glazing [J]. Renewable energy, 2016,92:450-461.

[112] K. A. R. Ismail,J. R. Henriquez. Simplified model for a ventilated glass window under forced air flow conditions [J]. Applied Thermal Engineering,2006,26:302-395.

[113] K. A. R. Ismail,J. R. Henriquez. Two-dimensional model for the double glass naturally ventilated window [J]. International journal of heat and mass transfer,2005,48:461-475.

[114] J. Xamán, Y. Olazo-Gómez, Y. Chávez,et al. Computational fluid dynamics for thermal evaluction of a room with a doube glazing window with a solar control film [J]. Renewable energy,2016, 94:237-250.

[115] J. A. Duffie,W. A. Beckman. *Solar Engineering of Thermal Processes*, *Third Edition* [M]. John Wiley & Sons,New York,2006.

[116] K. Y. A. Kondratyev. Radiation in the Atmosphere [M]. Academic press,Now York,1969.

[117] American Society of Heating,Refrigerating,Air Conditioning Engineers,ASHRAE Fundamentals [S]. Atlanta,GA. 2013.

[118] A. V. Rabadiya,R. Kirar. Comparative analysis of wind loss coefficient (wind heat transfer coefficient) for solar flat plate collector [J]. International Journal of Emerging Technology and Advanced Engineering,2012,2:463-468.

[119] W. A. Beckman. Solution of heat transfer problems on a digital computer [J]. Solar Energy, 1971,13:293-300.

[120] F. P. Incropera,D. P. Dewitt. *Fundamentals of Heat and Mass Transfer*,3rd ed [M]. Wiley,New YorK,1990.

[121] I. A. Macdonald. Quantifying the Effects of Uncertainty in Building Simulation [D]. Department

of Mechanical Engineering, University of Strathclyde, UK, 2002.

[122] C. Y. Li. Performance evaluation of water-flow window glazing [D]. Department of Civil and Architectural Engineering, City University of Hong Kong, Hong Kong SAR, China, 2012.

[123] T. P. Otanicar, P. E. Phelan, J. S. Golden. Optical properties of liquids for direct absorption solar thermal energy systems [J]. Solar Energy, 2009, 83: 969-977.

[124] A. Shukla, D. Buddhi, R. L. Sawhney. Solar water heaters with phase change material thermal energy storage medium: A review [J]. Renewable and Sustainable Energy Reviews, 2009, 13: 2119-2125.

[125] A. Abhat. Low temperature latent heat thermal energy storage: Heat storage materials [J]. Solar Energy, 1983, 30: 313-332.

[126] A. Sharma, V. V. Tyagi, C. R. Chen, et al. Review on thermal energy storage with phase change materials and application [J]. Renewable and Sustainable Energy Reviews, 2009, 13: 318-345.

[127] M. Mazman, L. F. Cabeza, H. Mehling, et al. Utilization of phase change materials in solar domestic hot water systems [J]. Renewable Energy, 2009, 34: 1639-1643.

[128] M. Medrano, M. O. Yilmaz, M. Nogués, et al. Experimental evaluation of commercial heat exchangers for use as PCM thermal storage systems [J]. Applied Energy, 2009, 86: 2047-2055.

[129] H. Shamai, M. Boroushaki, H. Geraei. Performance evaluation and optimization of encapsulated casade PCM thermal storage [J]. Journal of energy storage, 2017, 11: 64-75.

[130] M. Zera, Y. Kozak, V. Dubovsky, et al. Analysis and optimization of melting temperature span for a multiple-PCM latent heat thermal energy storage unit [J]. Applied thermal engineering, 2016, 93: 315-329.

[131] J. C. Kurnia, A. P. Sasmito, S. V. Jangam, et al. Improved design for heat transfer performance of a novel phase change material (PCM) thermal energy storage (TES) [J]. Applied thermal engineering, 2013, 50: 896-907.

[132] F. Agyenim, N. Hewitt, P. Eames, et al. A review of materials, heat transfer and phase change problem formulation for latent heat thermal energy storage systems (LHTESS) [J]. Renewable and Sustainable Energy Reviews, 2010, 14: 615-628.

[133] V. R. Voller. A fixed grid numerical modelling methodology for convection-diffusion mushy region phase-change problems [J]. Heat mass transfer, 1987, 30: 1709-1719.

[134] X. W. Cheng, X. Q. Zhai, R. Z. Wang. Thermal performance analysis of a packed bed cold storage unit using composite PCM capsules for high temperature solar cooling application [J]. Applied thermal engineering, 2016, 100: 247-255.

[135] F. Fornarelli, S. M. Camporeale, B. Fortunato, et al. CFD analysis of melting process in a shell-and-tube latent heat storage for concentrated solar power plants [J]. Applied Energy, 2016, 164: 711-722.

[136] K. C Zhong, S. H. Li, G. F. Sun, et al. Simulation study on dynamic heat transfer performance of PCM-filled glass window with different thermophysical parameters of phase change material [J]. Energy and Buildings, 2015, 106: 87-95.

[137] S. Almsater, A. Alemu, W. Saman, et al. Development and experimental validation of a CFD

model for PCM in a vertical triplex heat exchanger [J]. Applied thermal engineering, 2007, 116:344-354.

[138] J. Bony, S. Citherlet. Numerical model and experimental validation of heat storage with phase change materials [J]. Energy and Buildings, 2007, 39:1065-1072.

[139] E. Tumilowicz, C. L. Chan, P. W. Li, et al. An enthalpy formulation for thermocline with encapsulated PCM thermal storage and benchmark solution using the method of characteristics [J]. International Journal of Heat and Mass Transfer, 2014, 79:362-377.

[140] American Society of Heating, Refrigerating, Air Conditioning Engineers, ASHRAE HVAC Application [S], Atlanta, GA. 2007.

[141] V. R. Voller. Fast implicit finit-difference method for the analysis of phase change problems [J]. Numerical heat transfer, 1990, 17:155-169.

[142] A. D. Brent, V. R. Voller, K. J. Reid. Enthalpy-porosity technique for modelling convection-diffusion phase change: application to the melting of a pure metal [J]. Numerical heat transfer, 1988, 13:297-318.

[143] M. Lacroix. Numerical simulation of a shell-and-tube latent heat thermal energy storage unit [J]. Solar energy, 1993, 50:357-367.

[144] Cibse Guide, Volume B. The chartered institution of building services engineers [S]. London, 1986.

[145] M. Smyth, P. C. Eames, B. Norton. Evaluation of a freeze resistant integrated collector/storage solar water-heater for northern Europe [J]. Applied Energy, 2001, 68:265-274.

[146] B. A. Wilcox, C. S. Barnaby. Freeze protection for fiat-plate collectors using heating [J]. Solar Energy, 2011, 19:745-746.

[147] A. H. Abdullan, H. Z. Abou-Ziyan, A. A. Ghoneim. Thermal performance of flat plate solar collector using various arrangements of compound honeycomb [J]. Energy Conversion and Management, 2003, 44:3093-3112.

[148] N. D. Kaushika, K. Sumathy. Solar transparent insulation materials: A review [J]. Renewable and Sustainable Energy Reviews, 2003, 7:317-351.

[149] F. Zhou, J. Ji, J. Y. Cai, et al. Experimental and numerical study of the freezing process of flat-plate solar collector [J]. Applied Thermal Engineering, 2017, 118:773-784.

[150] Y. F. Li, W. K. Chow. Optimum insulation-thickness for thermal and freezing protection [J]. Applied Energy, 2005, 80:23-33.

[151] A. Liebold, D. Gstoehl, D. Oppliger, et al. Fluidglass-façade element for active solar control for high-rise buildings [R]. 3th international high performance buildings conference at purdue, Purdue, American. July 14-17, 2014.

[152] H. F. Liu, Y. Q. Jiang, Y. Yao. The field test and optimization of a solar assisted heat pump system for space heating in extremely cold area [J]. Sustainable Cities and Society, 2014, 13:97-104.

[153] H. F. Liu, S. C. Zhang, Y. Q. Jiang, et al. Feasibility study on a novel freeze protection strategy for solar heating systems in severely cold areas [J]. Solar Energy, 2014, 112:144-153.

[154] V. Trillat-Berdal, B. Souyri, G. Achard. Coupling of geothermal heat pumps with thermal solar collectors [J]. Applied Thermal Engineering, 2007, 27: 1750-1755.

[155] O. Ozgener, A. Hepbasli. Experimental investiation of the performance of a solar-assisted ground-source heat pump system for greenhouse heating [J]. International journal of energy research, 2005, 29: 217-231.

[156] B. Norton, J. E. J. Edmonds. Aqueous propylene-glycol concentrations for the freeze protection of thermosyphon solar energy water heaters [J]. Solar Energy, 1991, 47: 375-382.

[157] A. A. Mazen. Experimental study of thermal conductivity of ethylene glycol water mixtures [J]. European Journal of Scientific Research, 2010, 44 (2): 300-313.

[158] E. Shojaeizadeh, F. Veysi, T. Yousefi, et al. An experimental investigation on the efficiency of a Flat-plate solar collector with binary working fluid: A case study of propylene glycol (PG) water [J]. Experimental Thermal and Fluid Science, 2014, 53: 218-226.

[159] GB 50352-2005, Code for Design of Civil Buildings (in Chinese) [S].

[160] GB 50495-2009, Technical code for solar heating system (in Chinese) [S].

[161] D. Diakoulaki, A. Zervos, J. Sarafidis, et al. Cost benefit analysis for solar water heating systems [J]. Energy Conversion and Management, 2001, 42: 1727-1739.

[162] H. Cassard, P. Denholm, S. Ong. Technical and economic performance of residential solar water heating in the United States [J]. Renewable and Sustainable Energy Reviews, 2011, 15: 3789-3800.

[163] M. N. A. Hawlader, K. C. Ng, T. T. Chandratilleke, et al. Economic eval-uation of a solar water heating system [J]. Energy Conversion and Management, 1987, 27: 197-204.

[164] S. C. Sekhar, K. L. C. Toon. On the study of energy performance and life cycle cost of smart window [J]. Energy and Buildings, 1998, 28: 307-316.

[165] T. Mateus, A. C. Oliveira. Energy and economic analysis of an integrated solar absorption cooling and heating system in different building types and climates [J]. Applied Energy; 2009, 86: 949-957.

[166] E. Streicher, W. Heidemann, H. Müller-Steinhagen. Energy payback time——A key number for the assessment of thermal solar systems [J]. Proceedings of EuroSun, 2004, 20-23.

[167] A. Simons, S. K. Firth. Life-cycle assessment of a 100% solar fraction thermal supply to a European apartment building using water-based sensible heat storage [J]. Energy and Buildings, 2011, 43: 1231-1240.

[168] S. Kalogirou. Thermal performance, economic and environmental life cycle analysis of thermosiphon solar water heaters [J]. Solar Energy, 2009, 83: 39-48.

[169] S. A. Kalogirou. Environmental benefits of domestic solar energy systems [J]. Energy Conversion and Management, 2004, 45: 3075-3092.

[170] Y. Hang, M. Qu, F. Zhao. Economic and environmental life cycle analysis of solar hot water systems in the United States [J]. Energy and Buildings, 2012, 45: 181-188.

[171] R. Battisti, A. Corrado. Environmental assessment of solar thermal collectors with integrated water storage [J]. Journal of Cleaner Production, 2005, 13: 1295-1300.

[172] F. Ardente, G. Beccali, M. Cellura, et al. Life cycle assessment of a solar thermal collector [J]. Renewable Energy, 2005, 30 (7): 1031-1054.

[173] K. K. Tse. Liquid based photovoltaic/thermal (PV/T) co-generation system in real building application [D]. Department of Civil and Architectural Engineering, City University of Hong Kong, Hong Kong SAR, China, 2014.

[174] F. Ardente, G. Beccali, M. Cellura, et al. Life cycle assessment of a solar thermal collector: sensitivity analysis, energy and environmental balances [J]. Renewable Energy, 2005, 30 (2): 109-130.

[175] G. Weir, T. Muneer. Energy and environmental impact analysis of double-glazed windows [J]. Energy Conversation and Management, 1997, 39: 243-256.

[176] R. Y. Teenou. nergy and CO_2 emissions associated with the production of Multi-glazed windows [D]. Mid Sweden University, 2012.

[177] Hong Kong Electrical and Mechanical Services Department (EMSD), PBBEC_Guidelines_2007 [Z], 2007.

[178] Y. Q. Lu. Handbook for heating and air-conditioning design, 2nd edition (in Chinese), China Architecture & Building Press, Beijing, 2008.

[179] T. T. Chow, J. Ji. Environmental life-cycle analysis of hybrid solar photovoltaic/thermal systems for use in Hong Kong [J]. International Journal of Photoenergy, 2012, 1-9.